Wei Wei, MA, MLS
Editor

Scholarly Communication in Science and Engineering Research in Higher Education

Scholarly Communication in Science and Engineering Research in Higher Education has been co-published simultaneously as *Science & Technology Libraries*, Volume 22, Numbers 3/4 2002.

Pre-publication REVIEWS, COMMENTARIES, EVALUATIONS . . .

"**V**aluable. . . . The ideas in this book will undoubtedly be widely accepted as vital to the goal of better utilization of what is already available. The book also provides an excellent summary of the types of tools and systems still awaiting wider recognition. . . . A VALUABLE CONTRIBUTION TO THE STUDY OF SCIENCE AND ENGINEERING RESEARCH."

Ellis Mount, DLS
President
Mount Data Services
Editor
Sci-Tech Book News
Author
Mapped for Murder

Scholarly Communication in Science and Engineering Research in Higher Education

Scholarly Communication in Science and Engineering Research in Higher Education has been co-published simultaneously as *Science & Technology Libraries*, Volume 22, Numbers 3/4 2002.

Science & Technology Libraries Monographic "Separates"

Below is a list of "separates," which in serials librarianship means a special issue simultaneously published as a special journal issue or double-issue *and* as a "separate" hardbound monograph. (This is a format which we also call a "DocuSerial.")

"Separates" are published because specialized libraries or professionals may wish to purchase a specific thematic issue by itself in a format which can be separately cataloged and shelved, as opposed to purchasing the journal on an on-going basis. Faculty members may also more easily consider a "separate" for classroom adoption.

"Separates" are carefully classified separately with the major book jobbers so that the journal tie-in can be noted on new book order slips to avoid duplicate purchasing.

You may wish to visit Haworth's website at . . .

http://www.HaworthPress.com

. . . to search our online catalog for complete tables of contents of these separates and related publications.

You may also call 1-800-HAWORTH (outside US/Canada: 607-722-5857), or Fax: 1-800-895-0582 (outside US/Canada: 607-771-0012), or e-mail at:

docdelivery@haworthpress.com

Scholarly Communication in Science and Engineering Research in Higher Education, edited by Wei Wei, MA, MLS (Vol. 22, No. 3/4, 2002). *Examines science and technology libraries' difficulties with maintaining expensive journal subscriptions for both researchers and tenure-ready scholars; offers advice and examples of efficient improvements to make fact-finding and publication easier and more cost-efficient.*

Patent and Trademark Information: Uses and Perspectives, edited by Virginia Baldwin, MS, MLS (Vol. 22, No. 1/2, 2001). *"A lucid and in-depth presentation of key resources and information systems in this area." (Javed Mostafa, PhD, Victor H. Yngve Associate Professor, Indiana University, Bloomington)*

Information and the Professional Scientist and Engineer, edited by Virginia Baldwin, MS, MLS, and Julie Hallmark, PhD (Vol. 21, No. 3/4, 2001). *Covers information needs, information seeking, communication behavior, and information resources.*

Information Practice in Science and Technology: Evolving Challenges and New Directions, edited by Mary C. Schlembach, BS, MLS, CAS (Vol. 21, No. 1/2, 2001). *Shows how libraries are addressing new challenges and changes in today's publishing market, in interdisciplinary research areas, and in online access.*

Electronic Resources and Services in Sci-Tech Libraries, edited by Mary C. Schlembach, BS, MLS, and William H. Mischo, BA, MA (Vol. 20, No. 2/3, 2001). *Examines collection development, reference service, and information service in science and technology libraries.*

Engineering Libraries: Building Collections and Delivering Services, edited by Thomas W. Conkling, BS, MLS, and Linda R. Musser, BS, MS (Vol. 19, No. 3/4, 2001). *"Highly useful. The range of topics is broad, from collections to user services . . . most of the authors provide extra value by focusing on points of special interest. Of value to almost all librarians or information specialists in academic or special libraries, or as a supplementary text for graduate library courses." (Susan Davis Herring, MLS, PhD, Engineering Reference Librarian, M. Louis Salmon Library, University of Alabama, Huntsville)*

Electronic Expectations: Science Journals on the Web, by Tony Stankus, MLS (Vol. 18, No. 2/3, 1999). *Separates the hype about electronic journals from the realities that they will bring. This book provides a complete tutorial review of the literature that relates to the rise of electronic journals in the sciences and explores the many cost factors that may prevent electronic journals from becoming revolutionary in the research industry.*

Digital Libraries: Philosophies, Technical Design Considerations, and Example Scenarios, edited by David Stern (Vol. 17, No. 3/4, 1999). *"Digital Libraries: Philosophies, Technical Design Considerations, and Example Scenarios targets the general librarian population and does a good job of opening eyes to the impact that digital library projects are already having in our automated libraries." (Kimberly J. Parker, MILS, Electronic Publishing & Collections Librarian, Yale University Library)*

Sci/Tech Librarianship: Education and Training, edited by Julie Hallmark, PhD, and Ruth K. Seidman, MSLS (Vol. 17, No. 2, 1998). *"Insightful, informative, and right-on-the-mark. . . . This collection provides a much-needed view of the education of sci/tech librarians."* *(Michael R. Leach, AB, Director, Physics Research Library, Harvard University)*

Chemical Librarianship: Challenges and Opportunities, edited by Arleen N. Somerville (Vol. 16, No. 3/4, 1997). *"Presents a most satisfying collection of articles that will be of interest, first and foremost, to chemistry librarians, but also to science librarians working in other science disciplines within academic settings."* *(Barbara List, Director, Science and Engineering Libraries, Columbia University, New York, New York)*

History of Science and Technology: A Sampler of Centers and Collections of Distinction, edited by Cynthia Steinke, MS (Vol. 14, No. 4, 1995). *"A 'grand tour' of history of science and technology collections that is of great interest to scholars, students and librarians."* *(Jay K. Lucker, AB, MLS, Director of Libraries, Massachusetts Institute of Technology; Lecturer in Science and Technology, Simmons College, Graduate School of Library and Information Science)*

Instruction for Information Access in Sci-Tech Libraries, edited by Cynthia Steinke, MS (Vol. 14, No. 2, 1994). *"A refreshing mix of user education programs and contain[s] many examples of good practice."* *(Library Review and Reference Reviews)*

Scientific and Clinical Literature for the Decade of the Brain, edited by Tony Stankus, MLS (Vol. 13, No. 3/4, 1993). *"This format combined with selected book and journal title lists is very convenient for life science, social science, or general reference librarians/bibliographers who wish to review the area or get up to speed quickly."* *(Ruth Lewis, MLS, Biology Librarian, Washington University, St. Louis, Missouri)*

Sci-Tech Libraries of the Future, edited by Cynthia Steinke, MS (Vol. 12, No. 4 and Vol. 13, No. 1, 1993). *"Very timely. . . . Will be of interest to all libraries confronted with changes in technology, information formats, and user expectations."* *(LA Record)*

Science Librarianship at America's Liberal Arts Colleges: Working Librarians Tell Their Stories, edited by Tony Stankus, MLS (Vol. 12, No. 3, 1992). *"For those teetering on the tightrope between the needs and desires of science faculty and liberal arts librarianship, this book brings a sense of balance."* *(Teresa R. Faust, MLS, Science Reference Librarian, Wake Forest University)*

Biographies of Scientists for Sci-Tech Libraries: Adding Faces to the Facts, edited by Tony Stankus, MLS (Vol. 11, No. 4, 1992). *"A guide to biographies of scientists from a wide variety of scientific fields, identifying titles that reveal the personality of the biographee as well as contributions to his/her field."* *(Sci Tech Book News)*

Information Seeking and Communicating Behavior of Scientists and Engineers, edited by Cynthia Steinke, MS (Vol. 11, No. 3, 1991). *"Unequivocally recommended. . . . The subject is one of importance to most university libraries, which are actively engaged in addressing user needs as a framework for library services."* *(New Library World)*

Technology Transfer: The Role of the Sci-Tech Librarian, edited by Cynthia Steinke, MS (Vol. 11, No. 2, 1991). *"Educates the reader about the role of information professionals in the multifaceted technology transfer process."* *(Journal of Chemical Information and Computer Sciences)*

Electronic Information Systems in Sci-Tech Libraries, edited by Cynthia Steinke, MS (Vol. 11, No. 1, 1990). *"Serves to illustrate the possibilities for effective networking from any library/information facility to any other geographical point."* *(Library Journal)*

The Role of Trade Literature in Sci-Tech Libraries, edited by Ellis Mount, DLS (Vol. 10, No. 4, 1990). *"A highly useful resource to identify and discuss the subject of manufacturers' catalogs and their historical as well as practical value to the profession of librarianship. Dr. Mount has made an outstanding contribution."* *(Academic Library Book Review)*

Role of Standards in Sci-Tech Libraries, edited by Ellis Mount, DLS (Vol. 10, No. 3, 1990). *Required reading for any librarian who has been asked to identify standards and specifications.*

Relation of Sci-Tech Information to Environmental Studies, edited by Ellis Mount, DLS (Vol. 10, No. 2, 1990). *"A timely and important book that illustrates the nature and use of sci-tech information in relation to the environment."* *(The Bulletin of Science, Technology & Society)*

End-User Training for Sci-Tech Databases, edited by Ellis Mount, DLS (Vol. 10, No. 1, 1990). *"This is a timely publication for those of us involved in conducting online searches in special libraries where our users have a detailed knowledge of their subject areas."* *(Australian Library Review)*

Sci-Tech Archives and Manuscript Collections, edited by Ellis Mount, DLS (Vol. 9, No. 4, 1989). *Gain valuable information on the ways in which sci-tech archival material is being handled and preserved in various institutions and organizations.*

Collection Management in Sci-Tech Libraries, edited by Ellis Mount, DLS (Vol. 9, No. 3, 1989). *"An interesting and timely survey of current issues in collection management as they pertain to science and technology libraries." (Barbara A. List, AMLS, Coordinator of Collection Development, Science & Technology Research Center, and Editor, New Technical Books, The Research Libraries, New York Public Library)*

The Role of Conference Literature in Sci-Tech Libraries, edited by Ellis Mount, DLS (Vol. 9, No. 2, 1989). *"The volume constitutes a valuable overview of the issues posed for librarians and users by one of the most frustrating and yet important sources of scientific and technical information." (Australian Library Review)*

Adaptation of Turnkey Computer Systems in Sci-Tech Libraries, edited by Ellis Mount, DLS (Vol. 9, No. 1, 1989). *"Interesting and useful. . . . The book addresses the problems and benefits associated with the installation of a turnkey or ready-made computer system in a scientific or technical library." (Information Retrieval & Library Automation)*

Sci-Tech Libraries Serving Zoological Gardens, edited by Ellis Mount, DLS (Vol. 8, No. 4, 1989). *"Reviews the history and development of six major zoological garden libraries in the U.S." (Australian Library Review)*

Libraries Serving Science-Oriented and Vocational High Schools, edited by Ellis Mount, DLS (Vol. 8, No. 3, 1989). *A wealth of information on the special collections of science-oriented and vocational high schools, with a look at their services, students, activities, and problems.*

Sci-Tech Library Networks Within Organizations, edited by Ellis Mount, DLS (Vol. 8, No. 2, 1988). *Offers thorough descriptions of sci-tech library networks in which their members have a common sponsorship or ownership.*

One Hundred Years of Sci-Tech Libraries: A Brief History, edited by Ellis Mount, DLS (Vol. 8, No. 1, 1988). *"Should be read by all those considering, or who are already involved in, information retrieval, whether in Sci-tech libraries or others." (Library Resources & Technical Services)*

Alternative Careers in Sci-Tech Information Service, edited by Ellis Mount, DLS (Vol. 7, No. 4, 1987). *Here is an eye-opening look at alternative careers for professionals with a sci-tech background, including librarians, scientists, and engineers.*

Preservation and Conservation of Sci-Tech Materials, edited by Ellis Mount, DLS (Vol. 7, No. 3, 1987). *"This cleverly coordinated work is essential reading for library school students and practicing librarians. . . . Recommended reading." (Science Books and Films)*

Sci-Tech Libraries Serving Societies and Institutions, edited by Ellis Mount, DLS (Vol. 7, No. 2, 1987). *"Of most interest to special librarians, providing them with some insight into sci-tech libraries and their activities as well as a means of identifying specialized services and collections which may be of use to them." (Sci-Tech Libraries)*

Innovations in Planning Facilities for Sci-Tech Libraries, edited by Ellis Mount, DLS (Vol. 7, No. 1, 1986). *"Will prove invaluable to any librarian establishing a new library or contemplating expansion." (Australasian College Libraries)*

Role of Computers in Sci-Tech Libraries, edited by Ellis Mount, DLS (Vol. 6, No. 4, 1986). *"A very readable text. . . . I am including a number of the articles in the student reading list." (C. Bull, Kingstec Community College, Kentville, Nova Scotia, Canada)*

Weeding of Collections in Sci-Tech Libraries, edited by Ellis Mount, DLS (Vol. 6, No. 3, 1986). *"A useful publication. . . . Should be in every science and technology library." (Rivernia Library Review)*

Sci-Tech Libraries in Museums and Aquariums, edited by Ellis Mount, DLS (Vol. 6, No. 1/2, 1985). *"Useful to libraries in museums and aquariums for its descriptive and practical information." (The Association for Information Management)*

Data Manipulation in Sci-Tech Libraries, edited by Ellis Mount, DLS (Vol. 5, No. 4, 1985). *"Papers in this volume present evidence of the growing sophistication in the manipulation of data by information personnel." (Sci-Tech Book News)*

Role of Maps in Sci-Tech Libraries, edited by Ellis Mount, DLS (Vol. 5, No. 3, 1985). *Learn all about the acquisition of maps and the special problems of their storage and preservation in this insightful book.*

Fee-Based Services in Sci-Tech Libraries, edited by Ellis Mount, DLS (Vol. 5, No. 2, 1985). *"Highly recommended. Any librarian will find something of interest in this volume." (Australasian College Libraries)*

Serving End-Users in Sci-Tech Libraries, edited by Ellis Mount, DLS (Vol. 5, No. 1, 1984). *"Welcome and indeed interesting reading. . . . a useful acquisition for anyone starting out in one or more of the areas covered." (Australasian College Libraries)*

Management of Sci-Tech Libraries, edited by Ellis Mount, DLS (Vol. 4, No. 3/4, 1984). *Become better equipped to tackle difficult staffing, budgeting, and personnel challenges with this essential volume on managing different types of sci-tech libraries.*

Collection Development in Sci-Tech Libraries, edited by Ellis Mount, DLS (Vol. 4, No. 2, 1984). *"Well-written by authors who work in the field they are discussing. Should be of value to librarians whose collections cover a wide range of scientific and technical fields." (Library Acquisitions: Practice and Theory)*

Role of Serials in Sci-Tech Libraries, edited by Ellis Mount, DLS (Vol. 4, No. 1, 1983). *"Some interesting nuggets to offer dedicated serials librarians and users of scientific journal literature. . . . Outlines the direction of some major changes already occurring in scientific journal publishing and serials management." (Serials Review)*

Planning Facilities for Sci-Tech Libraries, edited by Ellis Mount, DLS (Vol. 3, No. 4, 1983). *"Will be of interest to special librarians who are contemplating the building of new facilities or the renovating and adaptation of existing facilities in the near future. . . . A useful manual based on actual experiences." (Sci-Tech News)*

Monographs in Sci-Tech Libraries, edited by Ellis Mount, DLS (Vol. 3, No. 3, 1983). *This insightful book addresses the present contributions monographs are making in sci-tech libraries as well as their probable role in the future.*

Role of Translations in Sci-Tech Libraries, edited by Ellis Mount, DLS (Vol. 3, No. 2, 1983). *"Good required reading in a course on special libraries in library school. It would also be useful to any librarian who handles the ordering of translations." (Sci-Tech News)*

Online versus Manual Searching in Sci-Tech Libraries, edited by Ellis Mount, DLS (Vol. 3, No. 1, 1982). *An authoritative volume that examines the role that manual searches play in academic, public, corporate, and hospital libraries.*

Document Delivery for Sci-Tech Libraries, edited by Ellis Mount, DLS (Vol. 2, No. 4, 1982). *Touches on important aspects of document delivery and the place each aspect holds in the overall scheme of things.*

Cataloging and Indexing for Sci-Tech Libraries, edited by Ellis Mount, DLS (Vol. 2, No. 3, 1982). *Diverse and authoritative views on the problems of cataloging and indexing in sci-tech libraries.*

Role of Patents in Sci-Tech Libraries, edited by Ellis Mount, DLS (Vol. 2, No. 2, 1982). *A fascinating look at the nature of patents and the complicated, ever-changing set of indexes and computerized databases devoted to facilitating the identification and retrieval of patents.*

Current Awareness Services in Sci-Tech Libraries, edited by Ellis Mount, DLS (Vol. 2, No. 1, 1982). *An interesting and comprehensive look at the many forms of current awareness services that sci-tech libraries offer.*

Role of Technical Reports in Sci-Tech Libraries, edited by Ellis Mount, DLS (Vol. 1, No. 4, 1982). *Recommended reading not only for science and technology librarians, this unique volume is specifically devoted to the analysis of problems, innovative practices, and advances relating to the control and servicing of technical reports.*

Training of Sci-Tech Librarians and Library Users, edited by Ellis Mount, DLS (Vol. 1, No. 3, 1981). *Here is a crucial overview of the current and future issues in the training of science and engineering librarians as well as instruction for users of these libraries.*

Networking in Sci-Tech Libraries and Information Centers, edited by Ellis Mount, DLS (Vol. 1, No. 2, 1981). *Here is an entire volume devoted to the topic of cooperative projects and library networks among sci-tech libraries.*

Planning for Online Search Service in Sci-Tech Libraries, edited by Ellis Mount, DLS (Vol. 1, No. 1, 1981). *Covers the most important issue to consider when planning for online search services.*

Scholarly Communication in Science and Engineering Research in Higher Education

Wei Wei
Editor

Scholarly Communication in Science and Engineering Research in Higher Education has been co-published simultaneously as *Science & Technology Libraries*, Volume 22, Numbers 3/4 2002.

The Haworth Information Press®
An Imprint of The Haworth Press, Inc.

New York • London • Victoria (AU)
www.HaworthPress.com

Q
223
.S235
2004

Published by

The Haworth Information Press®,10 Alice Street, Binghamton, NY 13904-1580 USA

The Haworth Information Press® is an imprint of The Haworth Press, Inc., 10 Alice Street, Binghamtom, NY 13904-1580 USA.

Scholarly Communication in Science and Engineering Research in Higher Education has been co-published simultaneously as *Science & Technology Libraries*™, Volume 22, Numbers 3/4 2002.

Cover design by Marylouise E. Doyle.

Library of Congress Cataloging-in-Publication Data

Scholarly communication in science and engineering research in higher education / [edited by] Wei Wei.
      p. cm.
    "Co-published simultaneously as Science & technology libraries, Volume 22, Numbers 3/4."
    Includes bibliographical references and index.
    ISBN 0-7890-2177-3 (hard cover : alk. paper) – ISBN 0-7890-2178-1 (soft cover : alk. paper)
    1. Communication in science. 2. Communication in engineering. I. Wei, Wei, 1952- . II. Science & technology libraries.
    Q223.S235 2004
    501'.4–dc22
                                     2003021893

Indexing, Abstracting & Website/Internet Coverage

This section provides you with a list of major indexing & abstracting services. That is to say, each service began covering this periodical during the year noted in the right column. Most Websites which are listed below have indicated that they will either post, disseminate, compile, archive, cite or alert their own Website users with research-based content from this work. (This list is as current as the copyright date of this publication.)

(continued)

(continued)

*Exact start date to come.

(continued)

Special Bibliographic Notes related to special journal issues
(separates) and indexing/abstracting:

- indexing/abstracting services in this list will also cover material in any "separate" that is co-published simultaneously with Haworth's special thematic journal issue or DocuSerial. Indexing/abstracting usually covers material at the article/chapter level.
- monographic co-editions are intended for either non-subscribers or libraries which intend to purchase a second copy for their circulating collections.
- monographic co-editions are reported to all jobbers/wholesalers/approval plans. The source journal is listed as the "series" to assist the prevention of duplicate purchasing in the same manner utilized for books-in-series.
- to facilitate user/access services all indexing/abstracting services are encouraged to utilize the co-indexing entry note indicated at the bottom of the first page of each article/chapter/contribution.
- this is intended to assist a library user of any reference tool (whether print, electronic, online, or CD-ROM) to locate the monographic version if the library has purchased this version but not a subscription to the source journal.
- individual articles/chapters in any Haworth publication are also available through the Haworth Document Delivery Service (HDDS).

Scholarly Communication in Science and Engineering Research in Higher Education

CONTENTS

ABOUT THE EDITOR

Wei Wei, MA, MLS, is Computer Science Librarian at the Science and Engineering Library, University of California, Santa Cruz, where she has also served as Library Instruction Coordinator, Reference Services Coordinator, and International Outreach Librarian. Ms. Wei is a member of the Special Libraries Association (SLA), the Science & Technology Division (SLA), and the Leadership and Library Management Division (SLA). She received the Impossible Award from the Science & Technology Division in 1995, and has written several articles for selected publications, including *Science and Technology News*, *Science & Technology Libraries*, the *Journal of Library & Information Science*, and the *Journal of Educational Media & Library Science*.

Introduction

Twelve articles following the theme of scholarly communication in science and engineering research in higher education are presented in this special volume. Our current state of scholarly publishing and communication is in crisis. The system of scholarly communication has broken down. Advanced technology and a restrained economy have made a great impact on the academic community and the way it communicates. In response to this crisis, many recent discussions concerning changes in the academic community and in scholarship, peer review and quality control, electronic journals and their costs, digital repository, digital archive and retrieval, as well as copyright issues are presented. As the academic community and libraries enter this rapidly changing networked environment, the articles in this thematic volume provide a variety of insights in these areas.

The traditional system of scholarly communication is undergoing changes. The vision of a future communication system of low cost, freely shared electronic exchange of scholarly information has withered to a hopeless fantasy. Calls for new initiatives of independent scholarly publication outside of the commercial mainstream are rising. The ability of the new technology and current digital experiments in scientific publishing have not yet solved the economic crisis in the academic community. Although there is much overlap among the issues discussed in the articles presented here, this volume is nonetheless divided into four separate but interrelated sections.

The first of six articles in the first section, Scholarly Publishing, "Can Peer Review Be Better Focused?" authored by Paul Ginsparg, discusses some important issues related to free access models in scientific publishing and looks in detail at the role of the arXiv and its implications for the tradi-

[Haworth co-indexing entry note]: "Introduction." Wei, Wei. Co-published simultaneously in *Science & Technology Libraries* (The Haworth Information Press, an imprint of The Haworth Press, Inc.) Vol. 22, No. 3/4, 2002, pp. 1-3; and: *Scholarly Communication in Science and Engineering Research in Higher Education* (ed: Wei Wei) The Haworth Information Press, an imprint of The Haworth Press, Inc., 2002, pp. 1-3. Single or multiple copies of this article are available for a fee from The Haworth Document Delivery Service [1-800-HAWORTH, 9:00 a.m. - 5:00 p.m. (EST). E-mail address: docdelivery@haworthpress.com].

http://www.haworthpress.com/store/product.asp?sku=J122
10.1300/J122v22n03_01

tional peer-review system. He suggests that advanced technology can provide efficient ways to access and navigate information, and at the same time, have more cost-effective means of authentication and quality control. "The Future of Scientific and Technical Journals," by Michel R. Dagenais, provides us with valuable insight on possible consequences of the transition from traditional print media to digital media, and discusses new roles of authors, reviewers, librarians and readers in his proposed efficient organization for future scholarly publishing. During the past years, many academic institutions initiated new programs or digital experiments in scientific publishing with the intention to free themselves from the traditional system of scholarly communication. One example of such a digital experiment is described in Catherine B. Soehner's article entitled "The eScholarship Repository: A University of California Response to the Scholarly Communication Crisis." Her article introduces the crisis in scholarly communication and presents solid reasons why the eScholarship Repository is an important experiment in the current landscape of scholarly publishing. Next are three articles addressing issues and concerns related to conference papers, theses, dissertations and journals in scientific publishing and their costs. According to her recent experience at Caltech, Kimberly Douglas suggests in her article, "Conference Proceedings at Publishing Crossroads," that conference papers in electronic publishing can increase dissemination, access and at the same time can significantly reduce cost. Susan Hall's "Electronic Theses and Dissertations: Enhancing Scholarly Communication and the Graduate Student Experience" outlines the importance of new initiative programs in electronic theses and dissertations as a form of scholarly communication and their authors, graduate students who are greatly impacted by this profound shifting. Dana L. Roth closes this section by examining "Chemistry Journals: Cost-Effectiveness, Seminal Titles and Exchange Rate Profiteering." He proposes a new cost-effectiveness metric calculated for several chemistry journals.

The second section, Scholarly Communication, begins with John Cruickshank's paper "The Role of Scientific Literature in Electronic Scholarly Communication." He provides an in-depth description of models of the scholarly communication circuit of scientists, the role of scientific literature in electronic scholarly communication and its impact on librarians and scholars. Kate Thomes in her "Scholarly Communication in Flux: Entrenchment and Opportunity," specifically looks at some activities that librarians may engage in to effect change at the local level in this ever changing scholarly communication environment.

The third section focuses on the topic of digital archive and retrieval. As digital technology has changed archives, questions remain how libraries can ensure that all digital material is preserved and stays accessible. Janet A. Hughes' article, "Issues and Concerns with the Archiving of Electronic Journals," particularly addresses the policies, problems and concerns of electronically archiving journals that have both print and online versions. One of the most important parts of examining this change is to look at how the digital media has changed expectations of users and how it affects the retrieving skills and users' information seeking behavior when conducting research. "User Expectations and the Complex Reality of Online Research Efforts," authored by David Stern, outlines the reality of current online research capabilities and discusses problems in using current online research tools.

Seldom looked at is the area of bibliometric analysis of citation data in electronic publishing and scholarly communication, which is the theme of the fourth section. Locke J. Morrisey emphasizes the importance of accurately linking together online publications and their references in his article, "Bibliometric and Bibliographic Analysis in an Era of Electronic Scholarly Communication." He concludes the article by suggesting a need for better bibliographic control, interactive systems as well as standard protocols for electronic publishing. On the other hand, Joseph R. Kraus examines the information seeking patterns of undergraduate students at the University of Denver in his article, "Citation Patterns of Advanced Undergraduate Students in Biology, 2000-2002." The findings of his recent study show that 76.2% of the citations cited by these students originate in journal articles, while only 1.0% of the citations originate from web sites.

It is my sincere hope that you find these articles stimulating and thought-provoking. In my introduction I mentioned the current "crisis" in scholastic communication. As often has been mentioned, the word "crisis" conveys the idea of an opportunity for the resolution of an issue or problem. In reviewing these timely articles I would emphasize that in the light of various recent technical innovations we are at the doorstep of applying significant enhancements to scholarly communication. We are fortunate as librarians and educators to have the opportunity to play an important role in improving the means and methods in this area of communication.

Wei Wei

SCHOLARLY PUBLISHING

Can Peer Review Be Better Focused?

Paul Ginsparg

SUMMARY. If we were to start from scratch today to design a quality-controlled archive and distribution system for scientific and technical information, it could take a very different form from what has evolved in the past decade from pre-existing print infrastructure. Recent technological advances could provide not only more efficient means of accessing and navigating the information, but also more cost-effective means of authentication and quality control. This article discusses relevant experiences of the past decade from open electronic distribution of research materials in

Paul Ginsparg, PhD, is Professor, Physics and Computer Science, and Faculty of Computing and Information, Cornell University, Ithaca, NY 14853 (E-mail: ginsparg@cornell.edu).

The author would like to thank David Mermin, Jean-Claude Guédon, Greg Kuperberg, Andrew Odlyzko, and Paul Houle for comments.

This text evolved from discussions originally with an American Physical Society publications oversight subcommittee on peer review, on which the author served in early 2002 along with Beverly Berger, Mark Riley and Katepalli Sreenivasan. © Paul Ginsparg. Printed with permission.

[Haworth co-indexing entry note]: "Can Peer Review Be Better Focused?" Ginsparg, Paul. Co-published simultaneously in *Science & Technology Libraries* (The Haworth Information Press, an imprint of The Haworth Press, Inc.) Vol. 22, No. 3/4, 2002, pp. 5-17; and: *Scholarly Communication in Science and Engineering Research in Higher Education* (ed: Wei Wei) The Haworth Information Press, an imprint of The Haworth Press, Inc., 2002, pp. 5-17. Single or multiple copies of this article are available for a fee from The Haworth Document Delivery Service [1-800-HAWORTH, 9:00 a.m. - 5:00 p.m. (EST). E-mail address: docdelivery@haworthpress.com].

http://www.haworthpress.com/store/product.asp?sku=J122
10.1300/J122v22n03_02

physics and related disciplines, and describes their implications for proposals to improve the implementation of peer review. *[Article copies available
for a fee from The Haworth Document Delivery Service: 1-800-HAWORTH. E-mail
address: <docdelivery@haworthpress.com> Website: <http://www.HaworthPress.com>.]*

KEYWORDS. arXiv, peer review, scholarly publishing, scientific publishing

FREE ACCESS MODELS

There has been much recent discussion of free access to the online
scholarly literature. It is argued that this material becomes that much
more valuable when freely accessible,[1] and moreover that it is in public
policy interests to make the results of publicly funded research freely
available as a public good.[2] It is also suggested that this could ultimately
lead to a more cost-efficient scholarly publication system. The response
of the publishing community has been that their editorial processes provide an essential service to the research community, that these are labor-intensive and hence costly, and that even if delayed, free access
could impair their ability to support these operations. (Or, in the case of
commercial publishers, reduce revenues to below the profit level necessary to satisfy their shareholders or investors.) Informal surveys (e.g.,[3])
of medium- to large-scale publishing operations suggest a wide range in
revenues per article published, from the order of $1000/article to more
than $10,000/article. The smaller numbers typically come from
non-profit operations that provide a roughly equivalent level of service,
and hence are more likely representative of actual cost associated to
peer reviewed publication. Even some of these latter operations are
more costly than might ultimately be necessary, due to the continued
need to support legacy print distribution, but the savings from eliminat-
ing print and going to an all-electronic in-house work-flow are estimated for a large non-profit publisher to be at most on the order of 30%.[4]
The majority of the expenses are for the non-automatable editorial oversight and production staff: labor expenses that are not only unaffected
by the new technology but that also increase faster than the overall in-
flation rate in developed countries.

A given journal could conceivably reduce its costs by choosing to
consider fewer articles, but this would not reduce costs in the system as
a whole, presuming the same articles would be considered elsewhere. If

a journal instead considers the same number of articles, but publishes fewer by reducing its acceptance rate, this results not only in an increased cost per published article for that journal, but also in an *increased* cost for the system as a whole, since the rejected articles resubmitted elsewhere will typically generate editorial costs at other journals. Moreover, in this case there is yet an additional hidden cost to the research community, in the form of redundant time spent by referees, time typically neither compensated nor accounted.

One proposal to continue funding the current peer-review editorial system is to move entirely from the subscription model to an "author-subsidy" model, in which authors or their institutions pay for the material, either when submitted or when accepted for publication, and the material is then made freely available to readers. While such a system may prove workable in the long-run, it is difficult to impress upon authors the near-term advantages of moving in that direction. From the institutional standpoint, it would also mean that institutions that produce a disproportionate amount of quality research would pay a greater *percentage* of the costs. Some could consider this unfair, though in the long-term a fully reformed and less expensive scholarly publication system should nonetheless offer real savings to those institutions, since they already carry the highest costs in the subscription model. Another short-term difficulty with implementing such a system is the global nature of the research enterprise, in which special dispensation might be needed to accommodate researchers in developing countries, operating on lower funding scales. Correcting this problem could entail some form of progressive charging scheme and a proportionate increase in the charges to authors in developed countries, increasing the psychological barrier to moving towards an author-subsidy system. (The other resolution to the problem of unequal resources–moving editorial operations to developing countries to take advantage of reduced labor costs–is probably not feasible, though it is conceivable that some of the production could be handled remotely.) A system in which editorial costs are truly compensated equitably would also involve a charge for manuscripts that are rejected (sometimes these require even more editorial time than those accepted), but implementing that is also logistically problematic.

The question for our scholarly research communications infrastructure is: if we were not burdened with the legacy print system and associated methodology, what system would we design for our scholarly communications infrastructure? Do the technological advances of the past decade suggest a new methodology that provides greater utility to the research enterprise at the same or lower cost?

CURRENT ROLES AND PERCEPTIONS

My own experience as a reader, author, and referee in Physics suggests that current peer review methodology in this field strives to fulfill roles for two different timescales: to provide a guide to expert readers (those well-versed in the discipline) in the short-term, and to provide a certification imprimatur for the long-term. But as I'll argue further below, the attempt to perform both functions in one step necessarily falls short on both timescales: too slow for the former, and not stringent enough for the latter. The considerations that follow here apply primarily to those many fields of research publication in which the author, reader, and referee communities essentially coincide. A slightly different discussion would apply for journal publication in which the reader community greatly outnumbers the author community, or vice versa.

Before considering modifications to the current peer review system, it's important to clarify its current role in providing publicity, prestige, and readership to authors. Outsiders to the system are sometimes surprised to learn that peer-reviewed journals do not certify correctness of research results. Their somewhat weaker evaluation is that an article is (a) not obviously wrong or incomplete, and (b) is potentially of interest to readers in the field. The peer review process is also not designed to detect fraud, or plagiarism, nor a number of associated problems–those are all left to posterity to correct. In many fields, journal publication dates are also used to stake intellectual property rights (indeed their original defining function).[5] But since the journals are not truly certifying correctness, alternate forms of public distribution that provide a trustworthy datestamp can equally serve this role.

When faculty members are polled formally or informally regarding peer review, the response is frequently along the lines "Yes, of course, we need it precisely as currently constituted because it provides a quality control system for the literature, signalling important contributions, and hence necessary for deciding job and grant allocations." But this conclusion relies on two very strong implicit assumptions: (a) that the necessary signal results directly from the peer review process itself, and (b) that the signal in question could *only* result from this process. The question is not whether we still need to facilitate *some* form of quality control on the literature; it is instead whether given the emergence of new technology and dissemination methods in the past decade, is the current implementation of peer review still the most effective and efficient means to provide the desired signal?

Appearance in the peer-reviewed journal literature certainly does not provide sufficient signal: otherwise there would be no need to supplement the publication record with detailed letters of recommendation and other measures of importance and influence. On the other hand, the detailed letters and citation analyses *would* be sufficient for the above purposes, even if applied to a literature that had not undergone that systematic first editorial pass through a peer review system. This exposes one of the hidden assumptions in the above: namely that peer-reviewed publication is a prerequisite to entry into a system that supports archival availability and other functions such as citation analysis. That situation is no longer necessarily the case. (Another historical argument for journal publication is that funding agencies require publication as a tangible result of research progress, but once again there are now alternate distribution mechanisms to make the results available, with other potential supplemental means of measuring impact.)

There is much concern about tampering with a system that has evolved over much of the past century, during which time it has served a variety of essential purposes. But the cat is already out of the bag: alternate electronic archive and distribution systems are already in operation, and others are under development. Moreover, library acquisition budgets are unable to keep pace even with the price increases from the non-profit sector. It is therefore both critical and timely to consider whether modifications of existing methodology can lead to a more functional or less costly system for research communication.

It is also useful to bear in mind that much of the current entrenched methodology is largely a post World War II construct, including both the large-scale entry of commercial publishers and the widespread use of peer review for mass production quality control. It is estimated that there are well over $8 billion/year in revenues in STM (Scientific, Technical, and Medical) primary publishing, for somewhere on the order of 1.5-2 million articles published/year. If non-profit operations had the capacity to handle the entirety, and if they could continue to operate in the $500-$1500 revenue per published article range, then with no other change in methodology there might be an immediate 75% savings in the system, releasing well over $5 billion globally. (While it is not likely that institutions supporting the current scholarly communications system would suddenly opt to reduce their overhead rates, at least their rate of increase might be slowed for a while, as the surplus is absorbed to support other necessary functions.) The commercial publishers stepped in to fulfill an essential role during the post World War II period, precisely because the non-profits did not have the requisite capacity to handle the dramatic increase in STM publishing with then-available technology. An altered methodology based on the electronic communications networks that evolved through the 1990s could prove better scalable to larger capacity. In this case, the technology of the 21st century would allow the traditional players from a century ago, namely the professional societies and institutional libraries, to return to their dominant role in support of the research enterprise.

arXiv ROLE AND LESSONS

The arXiv[6] is an automated distribution system for research articles, without the editorial operations associated to peer review. As a pure dissemination system, i.e., without peer review, it operates at a factor of 100 to 1000 times lower in cost than a conventionally peer-reviewed system.[3] This is the real lesson of the move to electronic formats and distribution: not that everything should somehow be free, but that with many of the production tasks automatable or off-loadable to the authors, the editorial costs will then dominate the costs of an unreviewed distribution system by many orders of magnitude. This is the subtle difference from the paper system, in which the expenses directly associated to print production and distribution were roughly the same order of magnitude as the editorial costs. When the two were comparable in cost, it wasn't as essential to ask whether the production and dissemination system should be decoupled from the intellectual authentication system. Now that the former may be feasible at a cost of less than 1% of the latter, the unavoidable question is whether the utility provided by the latter, in its naive extrapolation to electronic form, continues to justify the associated time and expense. Since many communities rely in an essential way on the structuring of the literature provided by the editorial process, a first related question is whether some hybrid methodology might provide all of the benefits of the current system, but for a cost somewhere in between the greater than $1000/article cost of current editorial methodology and the less than $10/article cost of a pure distribution system. A second question is whether a hybrid methodology might also be better optimized for the differing needs, on differing timescales, of expert readers on the one hand and neophytes on the other.

The arXiv was initiated in 1991, before any physics journals were online. Its original intent was not to supplant journals, but to provide equal and uniform global access to prepublication materials (originally it was only to have had a three month retention time). Due to the multi-year period from '91 until established journals did come online en masse, the arXiv de facto took on a much larger role, by providing the unique online platform for near-term (5-10 yr.) "archival" access. Electronic offerings have, of course, become commonplace since the early 1990s: many publishers now put new material online in e-first mode, and the searchability, internal reference linking, and viewable formats they provide are at least as good as those of the automated arXiv. These conventional publishers are also set up to provide superior services wherever manual oversight, at additional cost, can improve on the author's product: e.g., correcting bibliographic errors and standardizing the front- and back-matter for automated harvesting. (Some of these costs may ultimately decline or disappear, however, with a more standardized "next-generation" document format, and improved authoring tools to produce it–developments from which automated distribution systems will benefit equally.)

We can now consider the current roles of the arXiv and of the online physics journals and assess their overlap. Primarily, the arXiv provides instant pre-review dissemination, aggregated on a field-wide basis, a breadth far beyond the capacity of any one journal. The journals augment this with some measure of authentication of authors (they are who they claim to be), and a certain amount of quality control of the research content. This latter, as mentioned, provides at least the minimum certification of "not obviously incorrect, not obviously uninteresting"; and in many cases provides more than that, e.g., those journals known to have higher selectivity convey an additional measure of short-term prestige. Both the arXiv and the journals provide access to past materials; and one could argue that arXiv benefits in this regard from the post facto certification functions provided by the journals. It is occasionally argued that organized journals may be able to provide a greater degree of long-term archival stability, both in aggregate and for individual items, though looking a century or more into the future this is really difficult to project one way or another.

With conventional overlapping journals having made so much online progress, does there remain a continued role for the arXiv, or is it on the verge of obsolescence? Informal polls of physicists suggest that it remains unthinkable to discontinue the resource, that it would simply have to be reinvented because it plays some essential role not fulfilled by any other. Hard statistics substantiate this: over 20 million full-text downloads during calendar year '02, on average the full text of each submission downloaded over 300 times in the 7 years from '96-'02, and some downloaded in the tens of thousands of times. The usage is significantly higher than comparable online journals in the field, and, most importantly, the access numbers have accelerated upwards as the conventional journals have come online over the past seven years. This is not to suggest, however, that physicist users are in favor of rapid discontinuation of the conventional journal system either.

What then is so essential about the arXiv to its users? The immediate answer is: "Well, it's obvious. It gives instant communication, without having to wait a few months for the peer review process." Does that mean that one should then remove items after some fixed time period? The answer is still "No, it remains incredibly useful as a comprehensive archival aggregator," i.e., a place where for certain fields instead of reading any particular journal, or set of journals, one can browse or search and be certain that the relevant article is there, and if it's not there it's because it doesn't exist. (This latter archival usage is the more problematic with respect to the refereed journals, since the free availability could undercut the subscription-based financial models–presuming the author-provided version is functionally indistinguishable from the later journal version.)

It has been remarked[7] that physicists use the arXiv site and do not appear concerned that the papers on it are not refereed. The vast majority of submissions are nonetheless submitted in parallel to conventional journals (at no "cost" to the author), and those that aren't are most frequently items such as theses or contributions to conference proceedings that nonetheless have undergone some effective form of review. Moreover, the site has never been a random UseNet newsgroup-like free-for-all. From the outset, a variety of heuristic screening mechanisms have been in place to ensure insofar as possible that submissions are at least *of refereeable quality.* That means they satisfy the minimal criterion that they would not be peremptorily rejected by any competent journal editor as nutty, offensive, or otherwise manifestly inappropriate, and would instead at least in principle be suitable for review (i.e., without the risk of alienating or wasting the time of a referee, that essential unaccounted resource). These mechanisms are an important–if not essential–component of why readers find the site so useful: though the most recently submitted articles have not yet necessarily undergone formal review, the vast majority of the articles can, would, or do eventually satisfy editorial requirements somewhere. Virtually all are in that grey area of decidability, and virtually none are entirely useless to active physicists. That is probably why expert arXiv readers are eager and willing to navigate the raw deposited material, and greatly value the accelerated availability over the filtering and refinement provided by the journal editorial processes (even as little as a few months later).

A MORE FOCUSED SYSTEM

The idea of using prior electronic distribution to augment the referee process goes back at least to [8]. Proposals along the lines of decoupling peer review from arXiv distribution can be found in [9], and the notion of "overlay" journals is further discussed in [6,10]. A review of various "decoupling" and "author subsidy" models proposed in the mid to late 1990s, taking advantage of new technology to implement improvements in research communication, can be found in [11]. (Note, in particular, the "eprint moderator model,"[12] intended to reduce costs by reducing the amount of material distributed in a commercial manner.) Recent experience in physics and related disciplines continues to reinforce the desirability of experimentation within this model space, with the expectation that similar implementations will prove feasible in other disciplines.

According to the observations above, the role of refereeing may be over-applied at present, insofar as it puts all submissions above the minimal criterion through the same uniform filter. The observed behavior of expert readers indicates that they don't value that extra level of filtering above their preference for instant availability of material "of refereeable quality." Non-expert readers typically don't need the availability on the timescale of a few months, but do eventually need a much higher level of selective filtering than is provided on the short timescale. Expert readers as well could benefit on a longer timescale (say a year or longer) from more stringent selection criteria, for the simple reason that the literature of the past decade is always much larger than the "instantaneous" literature. More stringent criteria on the longer timescale would also aid significantly in the job and grant evaluation functions, for which signal on the year or more timescale remains sufficiently timely. More stringent evaluation could potentially play a far greater role than peer-reviewed publication currently does, as compared to external letters and citation analyses.

Can these considerations be translated into either a more functional or more cost-effective peer review system? As already discussed, editorial costs cannot be reduced by adopting a lower acceptance rate on some longer timescale, but with the same number of submissions considered as currently, and by the current methodology. Instead the simplest proposal is a two-tier system, in which on a first pass only some cursory examination or other pro forma certification is given for acceptance into a standard tier. This could be minimally labor-intensive, perhaps relying primarily on an automated check of author institutional affiliation, prior publication record, research grant status, or other related background; and involve human labor primarily to adjudicate incomplete or ambiguous results of an automated pass.

Then at some later point (which could vary from article to article, perhaps with no time limit), a much smaller set of articles would be selected for the full peer review process. The initial selection criteria for this smaller set could be any of a variety of impact measures, to be determined, and based explicitly on their prior widespread and systematic availability and citability: e.g., reader nomination or rating, citation impact, usage statistics, editorial selection. . . . The instructions to expert reviewers would be similar to those now, based on quality, originality, and significance of research, degree of pedagogy (for review articles), and so on. The objective would be greater efficiency by focusing the comprehensive process not only on a smaller subset, but also that with a higher likely acceptance rate. These are the articles most likely to be archivally useful, and hence merit the enhanced editorial treatment for upgrade into the upper tier, including, for example, text clarifications and other improvements. This would also reduce the inefficient expenditure of community intellectual resources on articles that may not prove as useful in the long-term. Upper tier enhancements could include anything from a thorough blind refereeing to open professional annotation and comment. The upper tier could also combine commentary on many related papers at once. The point is that it's possible to provide more signal of various sorts to users on a smaller subset of articles, without worry about fairness issues of limited dissemination for the rest, and this can be done at lower overall cost than the current system, both in time spent by editors and in elective time spent by referees.

The standard tier would provide a rapid distribution system only marginally less elite than much of the current publication system, and enormously useful to readers and authors. Articles needn't be removed from the standard tier, and could persist indefinitely in useful form (just as currently in the arXiv), available via search interfaces and for archival citation–in particular, they would remain no less useful than had they received some cursory glance from a referee. Rapid availability would also be useful for fields in which the time to publication time is perceived to be too long. The standard tier availability could also be used to collect confidential commentary from interested readers so that eventual referees would have access to a wealth of currently inaccessible information held by the community, and help to avoid duplication of effort. In addition, articles that garner little attention at first, or are rejected due to overly restrictive policies, only to be properly appreciated many years later, would not be lost in the short-term, and could receive better long-term treatment in this sort of system. Various gradations, e.g., appearance in conference proceedings, would also automatically appear in the standard tier and provide additional short-term signal occasionally useful to non-expert readers.

The precise criteria for entry into the standard tier would depend on its architecture. Adaptable criteria could apply if it is some federation of institutionally and disciplinarily held repositories. The institutional repositories could rely on some form of internal endorsement, while the disciplinary aggregates could rely either on affiliation or on prior established credentials ("career review"[13] as opposed to "peer review"). Alternate entry paths for new participants, such as referrals from prior credentialed participants or direct appeal for cursory editorial evaluation (not full-fledged peer review), would also be possible. The essential idea is to facilitate communication within the recognized research community, without excessive noise from the exterior.[9] While multiple logically independent (though potentially overlapping)[3] upper tiers could naturally evolve, only a single globally held standard tier is strictly necessary, with of course any necessary redundancy for full archival stability. Suitable licensing procedures or copyright retention[2] to facilitate such a system are consistent with the spirit of copyright law, "To promote the Progress of Science and useful Arts" (for a recent discussion, see[14]).

At the second stage, it might also be feasible and appropriate for the referees and/or editor to attach some associated text explaining the context of the article and the reasons for its importance. Expert opinion could be used not only to guide readers to the important papers on the subject, but also guide readers through them. This would be a generalization of review volumes, potentially more timely and more comprehensive. It could include both suggested linked paths through the literature to aid in understanding an article, and could also include links to subsequent major works and trends to which an article later contributed. This latter citation tree could be frozen at the time of the refereeing of the article, or could be maintained retroactively for the benefit of future readers. Such an overlay guide to the "primary" literature could ultimately be the most important publication function provided by professional societies. It might also provide the basis of the future financial model for the second stage review process, possibly a combination of subscription (electronic, or even print if desired) and upper tier author subsidy. It could subsidize part of the cost of the less selective "peer reviewable" designations in the first stage for those lacking institutional credentials, perhaps together with a first stage "editorial fee" far smaller than for the later full editorial process.

As just one partial existence proof for elements of this system, consider for example the Mathematical Reviews, published by the American Mathematical Society. It provides a comprehensive set of reviews of the entire mathematical literature and an invaluable resource to mathematicians. It currently considers on the order of 100,000 articles per year, and chooses to review approximately 55,000 of those, at a rough overall effective editorial cost of under $140 per review.[15] The expenses include a nominal payment to reviewers, and also curation and maintenance of historical bibliographic data for the discipline. (Mathematician Kuperberg has also commented that "Math Reviews and Zentralblatt are inherently more useful forms of peer review,"[16] though he observes ironically that their publishers do not share this conviction.) Mathematical Reviews uses as its information feed a canonical set of conventional mathematics journals. In the future, such an operation could conceivably use some equally canonicalized cross-section taken from a standard tier of federated institutional and disciplinary repositories, containing material certified to be "of peer reviewable quality." While not all upper tier systems need to aspire to such disciplinary comprehensivity, this does provide an indication that they can operate usefully at an order of magnitude lower cost than conventionally peer reviewed journals.

The modifications described here are intended as a starting point for discussion of how recent technological advances could be used to improve the implementation of peer review. They are not intended to be revolutionary, but sometimes a small adjustment, with seemingly limited conceptual content, can have an enormous effect. In addition, these modifications could be undertaken incrementally, with the upper tier created as an overlay on the current publication base, working in parallel with the current system. Nothing would be jeopardized, and any new system could undergo a detailed efficacy assessment that many current implementations of peer review have either evaded or failed.

NOTES

1. R. Stephen Berry, "Is electronic publishing being used in the best interests of science? The scientist's view," *Electronic Publishing in Science II*, UNESCO HQ, Paris, 2001 (eds. Sir R. Elliot and D. Shaw), http://users.ox.ac.uk/~icsuinfo/berryfin.htm.

2. Steven Bachrach, R. Stephen Berry, Martin Blume, Thomas von Foerster, Alexander Fowler, Paul Ginsparg, Stephen Heller, Neil Kestner, Andrew Odlyzko, Ann Okerson, Ron Wigington, Anne Moffat, "Who should own scientific papers?" *Science* 281: 1459-1460 (1998) http://www.sciencemag.org/cgi/content/full/281/5382/1459.

3. P. Ginsparg "Creating a Global Knowledge Network," in *Electronic Publishing in Science II*, proceedings of joint ICSU Press/UNESCO conference, Paris, 2001 (eds. Sir R. Elliot and D. Shaw), http://users.ox.ac.uk/~icsuinfo/ginspargfin.htm (copy at http://arXiv.org/blurb/pg01unesco.html).

4. This estimate is for the American Physical Society, which publishes over 14,000 articles per year, and derives from figures discussed with its publications oversight committee. The percentage estimated for other publishing operations will vary, especially when editorial time and overhead is differentially accounted. In the discussion that follows, however, it matters only that there will be no *windfall* savings to publishers from going all-electronic, while employing the same overall labor-intensive methodology.

5. Jean-Claude Guédon, "In Oldenburg's Long Shadow: Librarians, Research Scientists, Publishers, and the Control of Scientific Publishing," *Proceedings ARL Membership Meeting*, May 2001. http://www.arl.org/arl/proceedings/138/guedon.html.

6. See http://arXiv.org/. For general background, see P. Ginsparg "Winners and losers in the global research village," in *Electronic Publishing in Science I*, proceedings of joint ICSU Press/UNESCO conference, Paris, 1996 (eds. Sir R. Elliot and D. Shaw), http://users.ox.ac.uk/~icsuinfo/ginsparg.htm (copy at http://arXiv.org/blurb/pg96unesco.html).

7. "Brinkman Outlines Priorities, Challenges for APS in 2002," *APS News*, January 2002. http://www.aps.org/apsnews/0102/010208.html.

8. S. Rogers and C. Hurt, "How Scholarly Communication Should Work in the 21st Century," *The Chronicle of Higher Education*, October 18, A56 (1989).

9. P. Ginsparg, "First Steps Towards Electronic Research Communication," *Computers in Physics*, Vol. 8, No. 4, Jul/Aug 1994, p. 390; see also P. Ginsparg, "After Dinner Remarks," http://arXiv.org/blurb/pg14Oct94.html, presented at the APS e-print Workshop at LANL, 14-15 Oct 1994, http://publish.aps.org/EPRINT/KATHD/toc.html.

10. P. Ginsparg, "Los Alamos XXX," November 1996 *APS News Online*, http://www.aps.org/apsnews/1196/11718.html (copy at http://arXiv.org/blurb/sep96news.html).

11. Steven Gass, "Transforming Scientific Communication for the 21st Century," *Science & Technology Libraries*, vol. 19, no. 3/4, 2001, pages 3-18.

12. David Stern, "eprint Moderator Model," http://www.library.yale.edu/scilib/modmodexplain.html (version dated Jan 25, 1999).

13. Rob Kling, Lisa Spector, Geoff McKim, "Locally Controlled Scholarly Publishing via the Internet: The Guild Model," *The Journal of Electronic Publishing*, August, 2002. http://www.press.umich.edu/jep/08-01/kling.html.

14. John Willinsky, "Copyright Contradictions in Scholarly Publishing," *First Monday*, volume 7, number 11 (November 2002), http://firstmonday.org/issues/issue7_11/willinsky/index.html.

15. Private communications from past and current *Mathematical Reviews* editors Keith Dennis and Jane Kister, based on publicly available data.

16. Greg Kuperberg, "Scholarly mathematical communication at a crossroads," *arXiv:math.HO/0210144, Nieuw Arch. Wisk.* (5) 3 (2002), no. 3, 262-264.

The Future
of Scientific and Technical Journals

Michel R. Dagenais

SUMMARY. Scholarly publishing is undergoing a profound transition from the traditional print media to the electronic media. This transition is almost complete from the reader's viewpoint. Authors, reviewers and librarians have yet to reap the expected benefits of increased efficiency and reduced costs. This article examines the possible consequences of this change, surveys recent developments and proposes an efficient organisation for future scholarly publishing. The new role of authors, reviewers, librarians and readers in this proposed organisation is discussed. *[Article copies available for a fee from The Haworth Document Delivery Service: 1-800-HAWORTH. E-mail address: <docdelivery@haworthpress.com> Website: <http://www.HaworthPress. com> © 2002 by The Haworth Press, Inc. All rights reserved.]*

KEYWORDS. Electronic publishing, scholarly publishing, electronic journals, open archives, peer review

INTRODUCTION

While a few scientific journals were founded in the late 17th century, as mentioned in [Guedon01] [Sav99], the majority of active scientific

Michel R. Dagenais, PhD, is Professor, Department of Computer Engineering, Ecole Polytechnique de Montreal, P.O. Box 6079, Station Downtown, Montreal, Quebec, Canada H3C 3A7 (E-mail: michel.dagenais@polymtl.ca).

[Haworth co-indexing entry note]: "The Future of Scientific and Technical Journals." Dagenais, Michel R. Co-published simultaneously in *Science & Technology Libraries* (The Haworth Information Press, an imprint of The Haworth Press, Inc.) Vol. 22, No. 3/4, 2002, pp. 19-28; and: *Scholarly Communication in Science and Engineering Research in Higher Education* (ed: Wei Wei) The Haworth Information Press, an imprint of The Haworth Press, Inc., 2002, pp. 19-28. Single or multiple copies of this article are available for a fee from The Haworth Document Delivery Service [1-800-HAWORTH, 9:00 a.m. - 5:00 p.m. (EST). E-mail address: docdelivery@haworthpress.com].

10.1300/J122v22n03_03

communities established journals in the middle of the 19th century. The basic scholarly publishing organisation remained unchanged until the late 20th century, when a critical mass of scholarly authors and readers started to routinely access information through the Internet. The following sections review briefly the traditional scholarly publishing model, examine the impact of the new electronic media, survey recent developments in electronic publishing, propose a scenario for the future of scientific and technical journals and conclude with a discussion of the transition period.

HISTORIC ORGANISATION

Authors of scholarly publications do not get directly retributed for their work. Instead, the dissemination of their publications brings them recognition, funding and career advancement, and contributes to the progress of science through the exchange of ideas. Reviewers similarly perform their task benevolently but without tangible reward, other than having the satisfaction of doing their share to insure quality publications, being cited in the reviewers list and seeing the submitted material earlier than readers.

Once an article is submitted and reviewed, the editor must accept or reject it based on its expected relevance and importance to the readership, taking into account the *limits* imposed by the costs of typesetting, printing and shipping the accepted articles. Journals where technically sound articles are rejected because of cost issues are labeled here as *limited content* journals; suggesting to readers which articles are more important is a different issue and does not require *limiting* the content.

In these *limited content* journals, the editor decides the filtering criteria for the whole readership. A rejected article may be resubmitted iteratively by its author, until acceptance, to more appropriate journals (different topic, different scope, different acceptance criteria). This iterative process is lengthy and tedious for authors and causes the reviewers to collectively review the same paper several times. Readers also suffer from this iterative process. Articles are not visible until reviewed, sometimes several times, accepted, and finally placed in the publication queues caused by page quotas. The total delay is anywhere from six months to several years.

Readers most often access scholarly publications for free at their library. The librarians organize the scholarly journals to make these easily accessible to their members, negotiate prices with the publishers, and make representations to their institutions to obtain adequate funding. Publishers coordinate editorial boards and organize the production and dissemination of the printed scholarly journals. The publishers may be non-profit learned societies or for profit publishing houses.

THE ELECTRONIC MEDIA

Electronic publications differ from paper publications in two important ways. The delivery speed is perhaps the most striking difference. Also important is the relatively small cost of publishing an article available in electronic form [Odlyzko99], once it has been reviewed. This enables *full content* journals, as opposed to *limited content journals*, where all technically sound and relevant articles are included, freed from the *limits* imposed by cost related page quotas. The reader is not expected to read more articles each month, but rather to specify his personal relevance ranking in order to select articles to read.

The electronic media revolution was foreseen in the 1970s or even the 1960s but only started having a tangible impact when the proportion of networked researchers reached a critical mass in the late 1990s. During this period, the number of electronically accessible journals grew from a few to several thousands [Mog99]. Readers can now access articles electronically, printing instead of photocopying articles, directly from the publishers' sites. The primary tangible benefactors of this change are the readers who have a simpler access to more articles, and eventually the publishers, who can avoid some of the costs associated with printing. The librarians are somewhat less involved than previously.

Several changes have not yet happened, however. Authors and reviewers have not benefited much from the electronic media capabilities since *full content* journals are few. The cost to institutions is not diminishing either [ARL00] [Guedon01]. Furthermore, electronic subscriptions change the acquisition model to a one-year viewing right. If a library stops a subscription, it loses access to all electronic issues. With printed journals subscriptions, already acquired issues remained in the library collection.

It is interesting to examine how the electronic medium was used in other areas where fewer constraints were present. Indeed, news sites are informative examples [Netscape02] [Slashdot02]. They have an incentive to attract readers but are not involved in articles counting, publish or perish, pressure. While their paper based *limited content* counterparts reject a large number of genuinely interesting stories, based on the editor's judgement and on strict page quotas, *full content* electronic news sites carry all stories satisfying minimum requirements. Weights and topics are assigned to each story. The site then presents on its home page the top stories, while allowing individual readers to specify their own reading profile.

RECENT DEVELOPMENTS

With the accelerating growth in the number of electronically available journals, several publication models can be identified. The libraries at Yale and UC Davis universities, among others, are maintaining excellent lists of links on electronic publishing [Kopper02] [Stern02].

Pre-Print Archives

The pre-print archives are freely accessible Web repositories for newly written articles, without formal peer review. The arXiv.org [arxiv02] site contains over 200,000 articles. These archives alleviate the publication delays, and insure free accessibility to a certain extent; some authors remove their article once it is published in a journal. The issue of peer review, both for building author recognition and helping the readers, is not addressed.

These articles may be independently submitted to regular journals for peer review and publication. Most journals require unpublished material and copyright transfer, which is incompatible with pre-print archiving. Steven Harnad [Harnad01] and others have suggested signing modified copyright transfer agreements. If journals refuse the modified agreements, they suggest to consider the pre-print article as non-published or as a separate article. Harnad suggests to add, as a side note, the list of modifications brought to the article for its final journal version, thus keeping the articles distinct. It is likely that the journal publishers will enforce more restrictive conditions on submitted articles, to prevent copies to pre-print archives, as these archives become more popular.

Institutional Repositories

Institutional archives are typically established and maintained by the institution library. They archive all the institution scholarly publications and impose formatting, indexing and relevance standards. Articles and theses are submitted for archival by their authors (e.g., professors and students from the institution). Global indexing services across institutional repositories are offered by worldwide search engines dedicated to open scholarly archives [OAI01]. Several institutional repositories were established in 2002, including the Massachussets Institute of Technology DSpace [DSpace02], and the University of California eScholarship [eScholarship02].

The articles in a repository may also be submitted to journals for peer review. Since the dissemination is already insured by the institutional repositories and the global indexing services, some journals may simply consist of a table of contents, with pointers to articles in the individual repositories; they are often referred to as *overlay journals.*

Restricted Access Journals

Most publishers now provide electronic access to their journals in addition to printed copies. The current trend is to replace paper subscriptions by subscriptions to many more titles available electronically. This reduces significantly the cost per title but often results in slightly higher total cost [JISC02]. Moreover, the negotiation power of the libraries is weakened since stopping an electronic subscription means losing access to everything.

Readers in the institution get more convenient access to more titles and librarians avoid the indexing and storage of the paper copies. This, however, does not help the publication delay and iterative submission issues.

Freely Accessible Journals

Most freely accessible journals, like their restricted counterparts, are closely modeled after traditional paper journals. However, by requiring electronic correspondance and electronic article submission in publication ready format, they forego the clerical, typesetting and printing costs. It then becomes possible to operate a journal on a tiny budget, easily subsidized by institutions or learning societies, and thus, to offer free access.

Freely accessible journals often publish articles on their Web site as soon as accepted, as they are not bound by page quotas, publication queues and publication dates associated with the paper format. Some journals make articles available even before their review and acceptance or refusal. However, removing rejected articles is problematic as readers may have started to access and link to these articles. Several directories of free access journals are available, many of which are referenced in Charles W. Baily, "Scholarly Electronic Publishing Bibliography," 2002. See the URL listed under Bibliography [Baily02].

Federating Initiatives

The Open Archives Initiative (OAI) [OAI01] establishes metadata standards to insure that distributed, freely accessible archives can be globally indexed and searched. It is a concept similar to popular search engines (e.g., google.com) but more structured and specialized for scholarly publications archives. The underlying assumption is that institutions (universities, research laboratories) will establish articles and technical reports repositories. The GNU eprint software [EPrints02] is a ready-to-use software package to build an OAI compliant Web archive. It is already used by several institutions around the world. More recently, the CERN, where the World Wide Web originated, released its OAI compliant Document Server Software [CERN02]. A number of academic institutions have launched ambitious OAI compliant scholarly publications repositories [DSpace02] [eScholarship02] [PKP02].

The SPARC initiative [SPARC02], from the Association of Research Libraries, aims at creating low cost electronic journals to fight the spiraling costs of subscriptions. In recent years, after a number of acquisitions, a smaller number of large publishers gained control of a very significant proportion of the scholarly publishing market, reducing price competition. The SPARC strategy is evolving towards replacing traditional journals with institutional repositories coupled with different certification models, including overlay journals [Crow02].

PubMed Central is an archive of freely accessible journals in the field of Life Sciences [PubMed02]. It insures that all the articles from the participating journals remain permanently and freely accessible in a common, easily indexable, storage format. Most participants are freely accessible journals. Some restricted access journals, mostly published by non-profit societies, participate but submit their articles a few months after publication, to protect their subscription revenues.

The Public Library of Science [PLS02] is a non-profit organisation with many informal ties to PubMed Central. It is dedicated to making the world's scientific and medical literature freely accessible. As a first step, it circulated an open letter asking publishers to make scholarly publications freely available after a maximum of 6 months after publication. This letter, as of December 2002, was signed by over 30,000 researchers throughout the world. Relatively few publishers have followed their invitation. The second step was to set up new freely accessible scholarly publications in early 2003.

The Roquade repository project is particularly interesting [Sav99]. It is open to all researchers. It supports the simple conversion of traditional journals from print to electronic format but really aims at exploiting the interesting properties of the electronic medium by offering immediate access, and before and after publication peer reviews. Roquade is being replaced by a European project, FIGARO [FIGARO02], with funding from the European Community starting in January 2003.

THE FUTURE

Researchers finish a set of experiments and produce interesting results. They prepare an article using their favourite word processor with an appropriate stylesheet. An export filter allows producing a structured document (e.g., in XML DocBook format). The article is submitted to their institutional repository along with metadata (keywords, type of contribution, topic and subtopic . . .) and checked for valid structured format. One of the appointed moderators superficially checks the article within a day or so, as would be done for publishing technical reports, and makes it freely accessible in the Web repository.

The authors then submit their article to an electronic *full content* journal in the form of a pointer to their article in the institutional repository; researchers without an institutional repository would submit the article directly to the journal. The article is screened by a moderator, or may be directly accepted if coming from an accredited institutional repository. The article is made accessible on the journal site as unreviewed. Peer reviews are scheduled for the article.

Reviewers receive notification about the newly available article. After careful review, they validate the keywords and contribution type (research, empirical study, survey, tutorial) and provide a rating for different aspects of the article (originality, applicability, importance, presentation). These ratings and associated comments are attached anonymously to the article, which acquires the status *reviewed* when a quorum of reviewers have submitted their ratings. The authors may then quote the ratings of each of their publications in their *curriculum vitae*.

The authors may also send a pointer to their article for inclusion in an upcoming conference or workshop. The conference technical committee may use the article topics and ratings to decide its applicability to the conference.

Readers may go to the journal Web site at any time and perform ad hoc queries based on keywords, authors, dates, topics and ratings. They may also subscribe to receive e-mail notification of articles of interest (matching a specified query), at a specified interval (daily, weekly, monthly . . .). Readers may as well access global search engines which index all the open archives (institutional repositories and freely accessible journals).

These freely accessible *full content* electronic journals would operate on relatively small budgets since authors, reviewers and editors offer their work benevolently and all steps are fully automated. The real cost is thus the incremental bandwidth, storage space, backups and maintenance associated with the journal Web site. Even the storage requirements would be modest since journals would only contain the reviews and the pointers to the articles in the institutional repositories. These journals would be owned by non-profit organisations, typically learned societies, public research institutes, or universities.

CONCLUSION

The transition to electronic media is almost complete from the reader's viewpoint, but is still at the early adoption stage from the author's and reviewer's perspective. Readers access scholarly publications online, often from each author's informal online repository of published articles and reports. Online conference programs or abstracts are used as pointers to new material in a domain, indicating which titles and authors to look for in Web search engines. Journals are therefore seldom accessed; more, and more timely, material is available in the online repositories.

Yet, research libraries must still acquire, index and store the traditional journals, on top of setting up institutional repositories and offering support for accessing online versions of the traditional journals. Similarly, authors must still cope with iterative submissions as peer recognition and career advancement is still based on traditional journal publications.

This uneasy transition period could take several years since authors are likely to use all available exposure channels, librarians cannot neglect traditional information sources while they are still being used, and publishers, even non-profit learned societies, have no incentive to relinquish their control over scholarly publishing.

The solution to a quicker and less painful transition is, in the author's opinion, to provide an immediate and practical incentive for scholarly authors to use the new dissemination medium. Interim *limited content* electronic journals, closely modeled after traditional journals, lack the significant incentives provided by *full content* electronic journals. Scholars should group with other colleagues and their librarians and join one of the many scholarly publishing transformation associations [SPARC02] [PLS02] to participate in establishing these new freely accessible *full content* electronic journals, implemented as *overlay journals* linking to articles in institutional archives.

BIBLIOGRAPHY

[ARL00] Association of Research Libraries, "Monograph and Serial Costs in ARL Libraries, 1986-2000," 2000, http://www.arl.org/stats/arlstat/graphs/2000t2.html.

[arxiv02] *ArXiv.org e-Print archive*, 2002, http://www.arxiv.org.

[Baily02] Charles W. Baily, "Scholarly Electronic Publishing Bibliography," 2002, http://info.lib.uh.edu/sepb/sepb.html.

[CERN02] CDS Development Group, "CERN Document Server Software," 2002, http://cdsware.cern.ch/.

[Crow02] Raym Crow, "The Case for Institutional Repositories: A SPARC Position Paper," 2002, http://www.arl.org/sparc/IR/ir.html.

[DSpace02] *DSpace*, 2002, http://www.dspace.org/.

[EPrints02] *EPrints.org–Self-Archiving and Open Archives*, 2002, http://www.eprints.org/.

[eScholarship02] *eScholarship Repository*, 2002, http://escholarship.cdlib.org/.

[FIGARO02] *FIGARO*, 2002, http://www.figaro-europe.net.

[Guedon01] Jean-Claude Guédon, "Beyond Core Journals and Licenses: The Paths to Reform Scientific Publishing," 2001, http://www.arl.org/newsltr/218/guedon.html.

[Harnad01] Steven Harnad, "For Whom the Gate Tolls?" 2001, http://www.ecs.soton.ac.uk/~harnad/Tp/resolution.htm.

[JISC02] Research Support Libraries Group, "Final Report from the JISC Scholarly Communications Group to the Research Support Libraries Group," 2003, http://www.rslg.ac.uk/final/final.pdf.

[Kopper02] C. Kopper, "Scholarly Electronic Publishing Initiatives," 2002, http://www.lib.ucdavis.edu/healthsci/webpub.html.

[Mog99] Dru Mogge, "Seven Years of Tracking Electronic Publishing: The ARL Directory of Electronic Journals, Newsletters and Academic Discussion Lists," 1999, http://dsej.arl.org/dsej/2000/mogge.html.

[Netscape02] *Netscape*, 2002, http://www.netscape.com/.

[OAI01] *Open Archives Initiative*, 2001, http://www.openarchives.org/.

[Odlyzko99] A. M. Odlyzko, "The Economics of Electronic Journals," in *Technology and Scholarly Communication*, Ekman and R. E. Quandt editors, Univ. Calif. Press, pp. 380-393, 1999, http://www.dtc.umn.edu/~odlyzko/doc/economics.journals.ps.

[PKP02] *Public Knowledge Project*, 2002, http://pkp.ubc.ca/.

[PLS02] *Public Library of Science*, 2002, http://www.publiclibraryofscience.org/.

[PubMed02] *PubMed Central an Archive of Life Science Journals*, 2002, http://www.pubmedcentral.nih.gov/.

[Sav99] J. S. M. Savenije, N. J. Grygierczyk, "The Roquade Project: A Gradual Revolution in Academic Publishing," 1999, http://www.library.uu.nl/staff/savenije/publicaties/RoquadeProject.htm.

[Slashdot02] *Slashdot News for Nerds*, 2002, http://www.slashdot.org/.

[SPARC02] *The Scholarly Publishing and Academic Resources Coalition*, 2002, http://www.arl.org/sparc.

[Stern02] D. E. Stern, "Possible Journal Cost Solutions and Enhancements," 2002, http://www.library.yale.edu/scilib/jrnlsol.html.

The eScholarship Repository:
A University of California Response
to the Scholarly Communication Crisis

Catherine B. Soehner

SUMMARY. There is a perceived need to find an alternative method for publishing as a result of a growing scholarly communication crisis. In response, the University of California has developed the eScholarship Repository, which offers their faculty a central location for depositing working papers, pre-publication scholarship, and research reports. It provides persistent access and makes the content easily discoverable for anyone with Internet access. The repository is a viable alternative publishing mechanism, and is a tremendous boon to users who want quick and inexpensive access to a wide range of scholarly material. Its ability to stem the tide of increasing journal prices or provide a replacement to current publishing methods for faculty in their efforts to attain tenure and promotion, however, depends upon the establishment of a peer review system similar to what is currently found in traditional publishing. *[Article copies available for a fee from The Haworth Document Delivery Service: 1-800-HAWORTH. E-mail address: <docdelivery@haworthpress.com> Website: <http://www.HaworthPress.com> © 2002 by The Haworth Press, Inc. All rights reserved.]*

Catherine B. Soehner, BSN, MLS, is Head, Science & Engineering Library, University of California, Santa Cruz, 1156 High Street, Santa Cruz, CA 95064 (E-mail: soehner@ucsc.edu).

[Haworth co-indexing entry note]: "The eScholarship Repository: A University of California Response to the Scholarly Communication Crisis." Soehner, Catherine B. Co-published simultaneously in *Science & Technology Libraries* (The Haworth Information Press, an imprint of The Haworth Press, Inc.) Vol. 22, No. 3/4, 2002, pp. 29-37; and: *Scholarly Communication in Science and Engineering Research in Higher Education* (ed: Wei Wei) The Haworth Information Press, an imprint of The Haworth Press, Inc., 2002, pp. 29-37. Single or multiple copies of this article are available for a fee from The Haworth Document Delivery Service [1-800-HAWORTH, 9:00 a.m. - 5:00 p.m. (EST). E-mail address: docdelivery@haworthpress.com].

KEYWORDS. eScholarship, eScholarship Repository, institutional repository, University of California, scholarly communication crisis, alternative publishing method

INTRODUCTION

There is an impending, or some suggest, already existing crisis in scholarly communication.[1] In the literature, the crisis is often not specifically defined.[2] There is, however, a general understanding that the crisis arises from increasing journal prices, decreasing library budgets, and unchanging requirements to publish to attain tenure and promotion. For example, between 1986 to 2000, the price of journal subscriptions has risen 226 percent. In contrast, the Consumer Price Index increased just 49 percent over the same period of time.[3] As a result, serials cancellations have become routine, and librarians must spend more of their time deciding which journals to keep, both in print and online, to get the most bang for their buck. This leaves librarians with less time to deal with the scholarly communication crisis as a whole.[4]

This crisis is demonstrated most acutely in the science and engineering disciplines. Despite the initial promise of lower production costs due to electronic publishing, journal prices have increased many times over. The prices of databases that index journals have skyrocketed as well. Library budgets are taking hits across the country, and library funding is one of the first items to fall to an ever heavily wielded budgetary axe. Most importantly, the faculty requirement to "publish or perish" offers an ever decreasing benefit for the scholarly community at large because articles are being published in journals whose subscriptions are too expensive for university and college libraries to maintain.

As a result, there is a scholarly communication crisis–a situation in which researchers publish but fewer and fewer people are able to access the publications. This affects researchers in many ways. Scholars need to have access to the most recent and relevant research in their field, and rely on libraries to provide this access. When libraries cancel journal subscriptions, researchers are unable to keep current, and this will have a profound effect on the quality of current and future research. One tool that scholars (and university administrators) use to measure the importance and impact of research is the number of times a published article is later cited. If an article is published in a journal that is too expensive for most libraries to carry, subsequent citations to it will necessarily be limited, as will its concomitant impact. In sum, scholars are directly affected by the scholarly communication crisis because they rely on serials for teaching and research, and are required to publish in them to achieve tenure and promotion. They are involved in the crisis whether they know it or not.

Even though the greatest impact of the scholarly communication crisis seems to be on researchers, the scholarly community's understanding of it has been slow in coming. Only recently has there been an increased understanding of it outside the library walls.[5] The two parties most directly involved with this crisis–publishers and researchers– could potentially resolve this crisis between themselves; however, both sides seem too heavily invested in the status quo to effect real change. There have been some attempts to make scholarly communication a free public resource, such as the Public Library of Science (PLoS) initiative. Such endeavors may not succeed. PLoS, for example, met strong resistance from publishers, one of whom described it as a high risk venture.[6]

Chemical Abstracts and its online counterpart, SciFinder Scholar, provide a good example of what happens when researchers become publishers. The American Chemical Society accredits the chemistry departments of educational institutions, and also publishes Chemical Abstracts. To be accredited, the American Chemical Society requires that institutions subscribe to Chemical Abstracts or to SciFinder Scholar. Despite this obvious conflict, there have been no efforts made by researchers who are members of the American Chemical Society to change the status quo. As a result, all educational institutions with chemistry departments will maintain subscriptions to Chemical Abstracts regardless of the cost and no matter what other chemical journal subscriptions are cancelled because of lack of funds. This epitomizes the scholarly communication crisis. Even though scholars may find citations to useful literature using Chemical Abstracts/SciFinder Scholar, they are less likely to find the full-text of the cited articles in their libraries.

Libraries are a part of this scholarly communication crisis and are pressured by both scholars and publishers. Scholars often insist that libraries maintain and even expand current subscription levels, or may demand alternatives to current publishing options. Publishers market directly to librarians and entice libraries to maintain current subscription levels by "bundling" journals into a single product, which the publishers then offer at a discounted price. Finding themselves in the unenviable situation of serving two masters, libraries and librarians have risen to the occasion, and have developed innovative ways to address the scholarly communication crisis.

Some of the most promising library-driven projects include the ACLS History Project, BioOne, the Gutenberg-e History Project, Project MUSE, and Highwire Press, to name a few.[7] One of the most exciting recent innovations is the formation of institutional repositories. Institutional repositories are digital collections that capture and preserve the intellectual output of university communities.[8] "A growing number of authors are already self-posting their work, and institutional repositories refine this trend, building stable sustainable infrastructure to support global communication of faculty research," stated Herbert Van de Sompel, coordinator of digital library research at the Research Library of the Los Alamos National Laboratory and co-founder of the Open Archives Initiative.[9] One benefit of these institutional repositories, as envisioned by the Scholarly Publishing and Academic Resources Coalition (SPARC), is that they "offer an immediate complement to the existing scholarly publishing model while stimulating the emergence of new structures that will evolve over time, offering expanded benefits to institutions and scholars alike."[10]

THE eSCHOLARSHIP REPOSITORY
AT THE UNIVERSITY OF CALIFORNIA

The University of California created the California Digital Library (CDL) in 1997, as part of the Office of the President. The CDL was charged to provide a comprehensive system for the management of the University of California's shared digital collections and to use new technology to enhance sharing of the physical collections throughout the University.[11] The CDL, which opened to the public in January 1999, partners with the 10 University of California (UC) campuses, and is the consortial arm of the UC system. As a digital "co-library," complementing the physical libraries of the UC system, the CDL uses technology to efficiently share materials held by UC, to provide greater and easier access to digital content, and to join with researchers in developing new tools and innovations for scholarly communication.[12]

To meet its goals, the CDL created the "eScholarship" initiative in 1999. Hosted by CDL, eScholarship is dedicated to facilitating "scholar-led innovations in scholarly communication."[13] The overall goal of its activities is the development of an infrastructure for digitally-based scholarly communication that: facilitates the expressed mutual interests of UC, its faculty, and the broader scholarly community; leverages the formidable capabilities and strengths of UC in order to provide effective national leadership in this area; and supports and extends experimental reconfigurations of the components of scholarly communication by communities of scholars themselves.[14]

One of its most recent experiments, the eScholarship Repository, was launched in April 2002. The eScholarship Repository offers researchers and faculty a library managed and maintained central location for depositing any research or scholarly output deemed appropriate by their participating UC research unit, center or department. Appropriate submissions might include working papers, technical reports, previously published materials, and pre-publication scholarship.[15] The arXiv repository of high energy physics preprints, first developed in 1991, was the inspiration for this institutional working paper repository experiment.[16] The eScholarship Repository has the potential to be a viable alternative to commercial ventures or self-publishing by providing a highly visible, persistent platform for freely distributed research and by reducing the need for expensive commercial products of a similar nature.

The repository is basically a computer database in which UC researchers can electronically publish scholarly materials. The database is managed and maintained by eScholarship. CDL licenses software and technical support for the repository from the Berkeley Electronic Press (bepress), whose Web site is http://www.bepress.com. All material posted to the repository must be submitted through a participating campus unit. Submissions to and organization of the repository is controlled by these campus units directly through designated repository administrators. Once electronically published, these articles can be viewed by anyone with access to the Internet (with a few copyright restrictions, when applicable).

Eligibility to publish in the repository is not limited to faculty, but may include research staff, postdoctoral fellows, lecturers, students, and visiting faculty, among others. Authors outside of UC may participate if their work is approved by a participating UC campus unit. For example, a unit may use the repository to post papers from a conference they sponsored, which may include some UC authors and many from other institutions. The sponsoring unit may decide what is appropriate for their part of the repository. These units also have full editorial control over what they publish.

The key component to publishing in the repository is the UC scholarly group or unit. It is around these groups or units that the repository is organized. Units can consist of UC Departments, Organized Research Units (ORUs), Multi-campus Research Units (MRUs), Centers (which are sometimes subsidiaries of ORUs), Institutes and affiliates. The repository has developed an easy registration and training process that allows any UC unit to participate. Training requires about one to two hours and is available for participants by phone, or in a group session given on campus. Prior to training, repository agreements are signed by the unit head, and a repository administrator or administrators are identified for each unit. The agreement signed by the unit head or director is a guarantee that the unit will obtain certain assurances from the author.[17] In addition, when a unit signs up to participate, a questionnaire is provided that allows each unit to specify their policies and standards.[18] bepress then sets up each unit's site.

When an author wants their material posted to the repository, he or she submits it to the appropriate campus unit. If accepted by the unit the article is posted to the repository using bepress software and technology. Submission requires a web browser, and documents can be provided in Microsoft Word, RTF, PDF, or HTML formats. Articles submitted as Word or RTF files are automatically converted to PDF files by the bepress system.[19] Units can hoose, during setup, to allow for a final review by an author before posting. Once a paper is submitted, a citation to the paper will remain in the repository even if the author later requests that the paper be removed. In addition, authors may submit several different versions of a paper, and leave all of them visible; or they may request to have earlier versions of a paper removed, leaving only a citation to those earlier versions. The repository allows for the upload of associated content, i.e., pictures, spreadsheets, PowerPoint presentations, or other content along with the paper, using the same easy submission process. Authors retain the copyright for papers posted to the repository. The author agreement specifies a nonexclusive right to use.[20]

The repository can be searched by author, title, abstract, date, keyword, and other fields using Boolean operators. Full-text searches can be conducted on all papers except some legacy papers that were scanned as page images only and submitted as PDFs. The system also makes use of an alerting service. Users can sign up to receive e-mail notification when new publications are posted in their areas of interest. Since the repository is Open Archive Initiative (OAI) compliant, users can discover eScholarship Repository papers from OAI discover services such as OAIster (http://www.oaister.org/).

THE BENEFITS OF THE eSCHOLARSHIP REPOSITORY AND ITS FUTURE IMPACT

The main institutional benefit of the eScholarship Repository is increased visibility and use of materials created by UC scholars. Faculty in academia already use citation analysis to provide tangible evidence of the relevance and importance of their research. Once material is posted to the repository, it is immediately available to researchers around the world who have access to the Internet. Early adopters of the eScholarship Repository have already uploaded thousands of papers, and use data shows more than 10,000 full-text paper downloads within the first few months. Full text downloads increased to over 44,000 by December 2002. As use increases, the relevance of UC research can be demonstrated in a tangible way. It is also of great benefit to UC to have the scholarly works of others outside of the UC system readily available using the OAI-compliant search software.

Advantages to authors who submit their papers to the repository can be seen immediately. Posting material to the repository is easy using the bepress software. Once material is added to the repository, the author can keep track of the use of their work and can see how many people have visited the repository site of a particular campus unit. Other services, such as e-mail notification of new papers entering the system and fielded, and full-text searching, make retrieval of relevant materials easy and efficient. Increased visibility of the author's repository content seems promising in light of the use data noted above.

Libraries also benefit from the repository. Libraries at the ten UC campuses are not required to pay any technology costs because the bepress software has been licensed and the repository has been created and is centrally maintained for the entire UC system. Librarians are working with eScholarship to publicize the existence of the repository, which provides an avenue for faculty-librarian collaboration around a common goal. The SPARC white paper on institutional repositories notes, "as libraries move to support faculty digital publishing activities, the library's relevance to the faculty–and, consequently, the institution overall–will increase."[21]

CONCLUSION

There are two major interrelated components of the scholarly communication crisis: the high cost of journals and a continuing reliance on publishing for tenure/promotion. The eScholarship Repository offers an attractive alternative to commercial publishing and commercial working paper services. Posting material to the repository is free to authors, and libraries are not required to purchase the material produced by faculty on their campus or within the UC system. This means that it is less likely that important UC research will end up in high-priced journals the UC libraries are unable to purchase. Furthermore, all research entered into the repository is available worldwide for free and will remain available permanently.

Whereas the repository is a viable alternative publishing mechanism, its ability to provide a replacement publishing mechanism for faculty efforts to attain tenure and promotion depends greatly on the establishment of standards of quality for the content of the repository. The eScholarship Repository offers several methods for determining the quality of the materials submitted. The affiliation of the author with UC and/or the affiliation with a specific campus unit provides a level of quality by association. In the future, it is possible that similar departments from different campuses will create different standards for inclusion of scholarly materials. This could create competition to create the best repository site in the UC system, thereby providing another mechanism for review and the posting of high quality materials. Each unit's designated gatekeeper for repository submissions uses standards agreed upon by the unit. This assures a certain level of peer review which is more than simple self-publishing on the Web, but still below the level of the rigorous peer review offered by traditional publishing methods. An assistant professor working toward tenure and promotion could not possibly use the repository alone to make a case for advancement unless academia as a whole reassessed and revised the tenure/promotion process.

For now, the repository must settle for being an additional publishing method alongside the traditional publishing outlets currently available. As faculty and university administrations across the country become increasingly aware of the decreased availability of articles published in the pursuit of tenure and promotion, the impact of institutional repositories like the eScholarship Repository on scholarly communication will be greater. At that time, the values at the foundation of the eScholarship Repository, including no costs passed on to UC authors and libraries, a commitment to the permanent availability of materials submitted, and unrestricted worldwide accessibility may spark an even more radical change in publishing methods.

NOTES

1. " 'Scholarly communication' has been simply defined by Shaugnessy as the 'social phenomenon whereby intellectual and creative activity is transmitted from one scholar to another.' " In Patricia Milne, "Scholarly Communication: Crisis, Response and Future," *Australian Academic & Research Libraries* 30, no. 2 (1999): 70. "Scholarly communication can take many forms, but generally the term implies the use of serial publications." In a book review by Gaby Haddow, *The Australian Library Journal* 49, no. 3 (2000): 282. For the purpose of this article, scholarly communication will almost invariably mean scholarly publication in journals.

2. For example, in Patricia Milne, "Scholarly Communication: Crisis, Response and Future," *Australian Academic & Research Libraries* 30, no. 2 (1999): 70, the author talks about the "crisis" in quotes. In Andrew Richard Albanese, "Revolution or Evolution: Amid Numerous Models, Librarians Find They Have New Roles to Play in Shaping the Future of Scholarly Communication," *Library Journal* 126, no. 8 (2001): 48, the author refers to "an increased understanding of the crisis outside the library walls," and includes the following observation: "faculty are tired of hearing about the current crisis in journals or monographs or whatever," without ever directly discussing what the "crisis" specifically is.

3. Andrew Richard Albanese, "Revolution or Evolution: Amid Numerous Models, Librarians Find They Have New Roles to Play in Shaping the Future of Scholarly Communication," *Library Journal* 126, no. 8 (2001): 48.

4. Susan K. Martin, "ACRL Takes Up the Challenges of Scholarly Communication," *College & Research Libraries News* 63, no. 11 (2002): 786.

5. Albanese, *supra*, *Library Journal* 126, no. 8 (2001): 48.

6. *Ibid.*

7. *Ibid.*

8. Raymond Crow, *The Case for Institutional Repositories: A SPARC Position Paper*, 2002. Viewed on December 16, 2002, and available at http://www.arl.org/sparc/IR/ir.html.

9. "SPARC Issues Position Paper on Scholarly Communication," *Information Today* 19, no. 8 (2002): 40.

10. Rick Johnson, SPARC's enterprise director, quoted in "SPARC Issues Position Paper on Scholarly Communication," *Information Today* 19, no. 8 (2002): 40.

11. "Overview of the California Digital Library," viewed on December 27, 2002, and available at http://www.cdlib.org/about/overview.

12. "California Digital Library, Berkeley Electronic Press Announce Partnership for Scholarly Communication Initiatives," *Ascribe Higher Education News Service* (4 October 2001): 1-3. Viewed on December 12, 2002, and available at web6.infotrac.galegroup.com/itw/infomark/29/210/29700952w6/.

13. "About eScholarship," viewed on December 18, 2002, and available at http://escholarship.cdlib.org/about.html.

14. "History of eScholarship," viewed on January 17, 2003, and available at http://escholarship.cdlib.org/history.html.

15. "University of California eScholarship Repository," viewed on December 23, 2002, and available at http://repositories.cdlib.org/escholarship.

16. *Ibid.*

17. "Policies," viewed on January 17, 2003, and available at http://repositories.cdlib.org/escholarship/policies.html.

18. "Setup Questionnaire," viewed on January 17, 2003, and available at http://repositories.cdlib.org/escholarship/setup.html.

19. "Frequently Asked Questions about PDF Files," viewed on January 17, 2003, and available at http://repositories.cdlib.org/pdffaq.html.

20. "Policies," viewed on January 17, 2003, and available at http://repositories.cdlib.org/escholarship/policies.html.

21. Raymond Crow, *The Case for Institutional Repositories: A SPARC Position Paper*, 2002. Viewed on December 16, 2002, and available at http://www.arl.org/sparc/IR/ir.html.

Conference Proceedings
at Publishing Crossroads

Kimberly Douglas

SUMMARY. The potential intrinsic to electronic publishing provides conference conveners with the opportunity to position the papers presented to greater advantage of both authors and readers. Unfortunately, conference papers are being increasingly published in the most expensive vehicle, the formal peer-reviewed journal. This circumstance is counter-productive to the legitimate role of conference papers in scholarly communication. The experience at Caltech in electronically publishing the proceedings of an international conference shows that conference papers can be more effectively published online at significantly less cost, thus increasing dissemination and access. *[Article copies available for a fee from The Haworth Document Delivery Service: 1-800-HAWORTH. E-mail address: <docdelivery@haworthpress.com> Website: <http://www.HaworthPress.com> © 2002 by The Haworth Press, Inc. All rights reserved.]*

KEYWORDS. Conference proceedings, scholarly communication, publishing, peer-review, journal subscription costs, Elsevier, priority, E-Prints, digital repositories, Open Archives Initiative, electronic publishing, Caltech, cavitation

---

Kimberly Douglas, MA, MS, is Director, Sherman Fairchild Library of Engineering and Applied Science, and Manager of Technical Information Services, Caltech Library System (E-mail: kdouglas@caltech.edu).

Address correspondence to Kimberly Douglas, Sherman Fairchild Library, Mailcode 1-43, Caltech, Pasadena, CA 91125.

[Haworth co-indexing entry note]: "Conference Proceedings at Publishing Crossroads." Douglas, Kimberly. Co-published simultaneously in *Science & Technology Libraries* (The Haworth Information Press, an imprint of The Haworth Press, Inc.) Vol. 22, No. 3/4, 2002, pp. 39-50; and: *Scholarly Communication in Science and Engineering Research in Higher Education* (ed: Wei Wei) The Haworth Information Press, an imprint of The Haworth Press, Inc., 2002, pp. 39-50. Single or multiple copies of this article are available for a fee from The Haworth Document Delivery Service [1-800-HAWORTH, 9:00 a.m. - 5:00 p.m. (EST). E-mail address: docdelivery@haworthpress.com].

10.1300/J122v22n03_05

ROLE OF PROCEEDINGS
IN SCHOLARLY COMMUNICATION

Scholarly communication consists of a multitude of vehicles to achieve different purposes. There are peer-reviewed articles; there are letters and notes; there are technical reports and preprints and there are conference papers. So much attention has been paid to the peer-reviewed article genre of late that any discussion regarding the dissemination of conference proceedings papers has electronically been nearly drowned out. Yet, as sci-tech librarians know, conference papers remain an integral part of the scholarly communication process. After all, Malinowsky wrote "Papers . . . at . . . meetings are original; very often they formulate hypotheses and syntheses of the first order of importance. Thus . . . they constitute primary sources. They are perhaps not of equal significance to the periodical article or the technical report, but still their import cannot be denied."[1]

The fundamental purpose of conferences is the exchange of new research results with the opportunity for immediate peer input. There's no substitute for a conference to energize creative thinking and new research approaches. Specialists and students gather together and create synergy, a natural result of direct human interaction. Tangible excitement of discovery, and learning and unfettered exchange of ideas are the lifeblood of conferences and no active researcher can maintain momentum or energy without such participation. Indeed, scientists are known to bemoan the proliferation of conferences while at the same time recognizing the proceedings as valuable.[2] In his paper ten years ago, Barschall appeals to his colleagues to hold fewer conferences so that resources could be applied in a more concentrated fashion. Fewer conferences would lead to higher attendance and fewer proceedings to publish would reduce costs while increasing coverage for libraries.

That was ten years ago and the problem remains. In fact Barschall's parting recommendation is that organizers of conferences "will examine the need for publishing proceedings and that, if they do wish to publish proceedings, they won't publish them in a journal unless libraries have the option not to purchase the proceedings."[3] Quite the opposite has occurred. The inclusion of conference papers in pre-paid journal subscriptions is becoming the rule. Allen[4] describes the vicious circle very clearly. Science libraries have less discretionary funding for books or any singly purchased items because the serials or journals budget commands an ever-increasing portion of the available funds, particularly in science libraries. Libraries purchase fewer monographic proceedings, and publishers respond by including proceedings in the journals to reach a guaranteed market, much larger than the conference on its own might attract. Conferences also provide an automatic theme, probably a "hot" one at that, for the journal, which is another marketing plus.[5] While it may be a good business decision and makes money for the publisher, it is not good for the information exchange necessary for quality research.

ACCESS TO PROCEEDINGS IN LIBRARY CATALOGS

It has been policy for many years at Caltech to add entries[6] or "analytics" to the library's catalog for conferences that appear in research journals. Conferences have been of sufficient interest to the Caltech community to warrant this extra work for two reasons: (1) conferences generally remain illusive and difficult to track down and (2) indexing services were not reliable in describing the conference content of a journal. In addition Caltech adds a fixed field code to cataloging records to indicate whether the item described is a conference publication. This is done primarily to aid discovery in the online catalog. Journals at Caltech are housed roughly by subject orientation: biology, chemistry, physics, math, and engineering creating a workable subject breakdown for studies of this kind. While Caltech is not a member of the Association of Research Libraries, it is most certainly a major research university for the sciences and its journal collection responsibly and consistently reflects the publishing record in those areas.

When a study (see Figure 1) of the presence of conference proceedings in the Caltech Libraries collection over the last ten years was conducted, we found that the publication of conferences in the science and engineering journal issues had grown 221% in the years from 1991 through 2001. This increase has occurred most severely in the subject areas of the most expensive journals and where the commercial publishers have the greatest presence, namely biology and chemistry. The greatest growth has been in chemistry with a change of 450%.

PROCEEDINGS AND PEER-REVIEW

This is a rather startling revelation since publishers and authors alike argue that the peer-reviewed journal is the sacred bearer of the highest quality papers.[7] Again Barschall commented that "a referee report on a

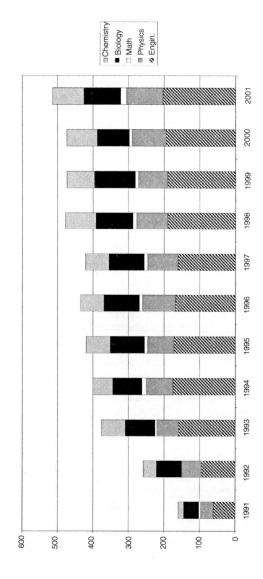

FIGURE 1. Number of Conferences Published in Research Journals Held at the Caltech Libraries

42

conference contribution is a rare occurrence . . . Conference proceedings are intended to be a record of what was actually presented at the conference. Hence altering a paper in response to a referee report, while desirable for an ordinary journal article, is not really appropriate for a paper in conference proceedings."[8] Tagler at Elsevier even commented that not all papers at a conference are necessarily published.[9]

Chemistry librarians have been particularly active in questioning the role of conference papers inside the covers of peer-review journals.[10] This issue has primarily arisen due to the extraordinary costs of those journals.[11] No consumer can continue to absorb double-digit annual price increases without investigating the reasons for the increase or the value of the product. The issue of conference papers appearing in peer-reviewed journals is certainly pertinent to both.

The exchange between the chemistry librarians and the editors of the *Journal of Molecular Structure*[12] regarding the appearance of conference papers in that journal and the added comments by the publisher[13] is notable for the emphasis by all parties on "peer-review." The librarians claim that conference papers are not peer-reviewed and the editors and publisher claim they are. The librarians use impact factor analysis as an attempt to apply an objective measure. The editors claim it's irrelevant because they know their community. Who can properly evaluate? For the International Congress on Peer Review in Biomedical Publication, Drummond Rennie notes in 2002 that what he described in 1986 as a system with "scarcely any bars to eventual publication" continues to be true.[14] And that is from a group that is proactively working on open peer-review standards. The situation in other disciplines lacking a comprehensive and systematic review of the review methodology cannot be much better. One might reasonably conclude that though there is a consensus that "peer-review" is a quality criteria, it exists primarily as an ideal, not a quantifiable measure. Business needs can easily and quickly trump any ideal.[15]

Librarians and authors both know that any single research result is described and communicated through multiple vehicles. The general understanding has been that a published conference paper is, in fact, the paper presented at the conference.[16] Surely that is the expectation of the readers and the librarians. After a conference presentation and ensuing Q and A, the author reworks the paper for submission to a peer-reviewed journal. The resulting published paper then reflects the value-add of peer critique and editorial treatment. The conference paper gets the word out; it forms the basis for a discussion. The peer-reviewed paper is the matured and tested description and analysis.

The most benign interpretation of this trend to publish conference papers in journals is to interpret it as yet another convergence phenomenon in the electronic information age. In order to make the content of a published journal compete with pre-print servers, news journals, and author websites, the publisher goes after the conference papers. Inclusion of new information in as timely a manner as possible (albeit after nearly a year delay in some cases) also puts the journal on the leading "news" edge. In turn the authors can show peer-reviewed papers for their work. Allen speculated that the necessity of rapid publishing "might be the driving force behind less rigorous reviewing of conference-derived papers that appear in journals."[17] One might ask, "Do proceedings, formally published in print, continue to serve the research community?" They do not. The primary reasons are: (1) excessive cost which creates access barriers and (2) a significant delay in distribution, which is contrary to the whole purpose of a conference.

Interestingly, since authors and publishers alike vigorously support adherence to traditional scholarly publishing (e.g., the peer-reviewed article journal) should they not then also logically argue that conference papers be published ad hoc, on the Web *before* the conference convenes? This would make a clear distinction between the conference paper and the peer-reviewed paper. It would establish priority for the author with probably the earliest time-stamp.

The exchange and sharing of research results is critical to the scientific method and is a recognized core value.[18] Certainly since Robert Merton's studies in the forties,[19] this core principle in research has been formally acknowledged. We also know that historically, research journals began as an exchange commodity between mutually interested organizations. This "exchange" behavior continued formally in research libraries well into the 1970s and remains alive to this day within well-defined small communities of interest.

In 2001 the Caltech Libraries had the opportunity to publish the contributed papers for an international conference that was held on the Pasadena campus. The conference convener, a professor of Mechanical Engineering, required that the "publishing" be electronic and that the papers be available online before the conference dates. The object was to make the most of participants' time by creating a convenient and easy to use website and to maximize distribution of the authors' work. At the same time costs were to be kept to a minimum. Certainly printing and mailing of paper products would be eliminated.

The Caltech Library System was already developing digital collections within the Open Archives Initiative compliant system, E-Prints, developed by Steven Harnad's group at the University of Southhampton in Great Britain. The conference papers were a first excursion into creating a "born" digital repository.

Cavitation 2001(http://cav2001.caltech.edu)[20] included 110 papers submitted from 18 different countries worldwide. Authors came from Asia, Eastern Europe and the Middle East. Though small, this conference represented the full spectrum of Internet access and use around the world. The conference conveners produced most of the publicity for the conference by e-mail and on the Internet. As preparations proceeded, linked web pages were added, enriching the main site located at another university, University of Michigan. Authors submitted abstracts to the conveners who reviewed them by standard e-mail methods. Eventually a list of approved papers and abstracts was prepared and submitted to the library staff, who then invited the authors to submit their papers directly into the digital library system. Once submitted, library staff reviewed the format and useability of the document's format and added or edited necessary metadata before final acceptance and release for general use. About 10% of the papers required resubmission from the author due to formatting issues. Most of the papers were submitted in pdf format (65%); another 30% were submitted in pure LaTeX that had to be rendered in pdf; and a few papers were submitted in MS Word. While there were a few formatting problems, overall any difficulties that arose were far fewer than expected. The authors were clearly conversant in preparing their documents. A few authors had network bandwidth problems and resorted to e-mail attachments as the method of submitting their papers. Work on creating the repository began in April 2001 and by the end of May, nearly all papers had been submitted.

Library staff spent nearly 200 hours on this first time effort, which included a certain amount of trial-and-error learning. Journal publishers report that a peer-review paper will generally cost at least $1,500 to produce. In some cases of purely electronic journals, it is said to cost $500/paper. Paul Ginsparg[21] points out that such costs are not sustainable. There isn't enough funding in academia to continue to pay these and even higher costs for certification, much less publishing and distribution of an ever-increasing body of literature. The Caltech experience showed that it was definitely possible to generate useable and accessible material at a much lower cost. The Caltech costs were under $100 per paper; a cost that would be reduced by at least one-half in a repeat effort. Enhancements to the E-prints software would further eliminate more of the currently necessary human interventions.

One might argue that the *CAV 2001* proceedings online is limited to the presentation format only, and therefore has not risen to the level of sophistication required for guaranteed archiving. Though the first statement is true, the environment is in constant flux and archiving solutions will be implemented as they become more standard and available. For the purposes of this paper it is important to recognize that the material was up, useable and ready for the participants before the conference and the content remains there now, available to the entire world.

Now consider the return on investment. Even now, 18 months after the conference was held, there are over 7,000 human hits on the conference site per month as measured by pdf downloads from web browsers that comprise 87% of the total accesses over this time period (see Figure 2). Contrary to other experiences, web browsers far outstrip the accesses by web crawlers and other harvesting 'bots. The number after initially falling off from a peak of about 11,000 during June 2001 to 6,000 in the summer months of 2001 continues to grow much to everyone's amazement.[22] Without a doubt, the group of researchers who participated in this conference will not conduct their conference without mounting the papers openly on the Web. The success has been overwhelming.

A look at the access logs reveals that many industrial sites, military facilities and education organizations around the world are making regular use of these papers. This stands to reason. Cavitation is a fluid mechanics phenomenon in which water flowing at high speeds creates bubbles that in bursting against surfaces pit and damage adjacent surfaces. Turbine and pump designs must address this engineering challenge. The invited lectures, informative review papers for the most part, are the most visited. An important note is that many of these site visitors are from organizations that are unlikely to have license agreements for access with major publishers. Given that this conference was the fourth in a series and that the earlier three conferences are relatively unknown, this usage speaks volumes for the benefit that unfettered access to conference papers allows. Depending on a researcher's needs the ready access to a conference paper could, in fact, drive more interest in the peer-reviewed version. For authors who want to have their work recognized and used, there is simply no comparison between an article openly available on the Internet and another that is closely controlled by a publisher, as authors are beginning to notice.

It is incomprehensible that in this day and age, with the maturity of the Web and the increasing ubiquity of access, that conference papers would continue to be printed at all, much less primarily in the formal journal lit-

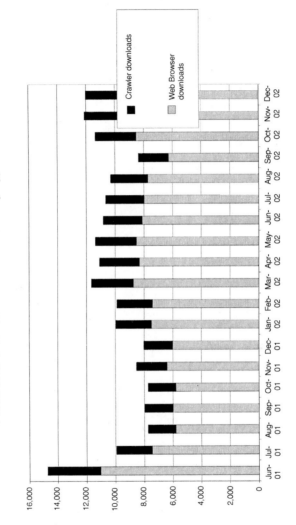

FIGURE 2. CAV2001 PDF Downloads by Type of Access

erature. The web is the perfect place for conference papers. They can be viewed by everyone; they can be openly commented on. They create documented attribution establishing priority for authors. They provide an opportunity for any conference convener to "use the Web to its full potential";[23] to enhance scholarly communication world-wide, all in preparation for the formal paper in peer-review journals.

NOTES

1. H. Robert Malinowsky, Richard A. Gray, Dorothy A. Gray, *Science and Engineering Literature*, 2nd ed. (Littleton, CO: Libraries Unlimited, Inc., 1976). p. 30.

2. Henry H. Barschall, Willy Haeberli, "What's Wrong with These Conferences?" *Physics Today*, no. 45 December (1992). p. 79.

3. Ibid. p. 81.

4. Robert S. Allen, "The Magnitude of Conference Proceedings in Physics Journals," *Special Libraries*, no. 86 Spring (1995). p. 142.

5. Ibid. p. 143. Allen points out the very real fair-use problems with this kind of packaging. Whole issues cannot easily be borrowed or otherwise acquired through interlibrary loan.

6. Dana Roth, "Extending the Online Catalog," in *Questions and Answers: Strategies for Using the Electronic Reference Collection, Clinic on Library Applications of Data Processing* (Urbana-Champaign: Univ. of Illinois, 1987). p. 36.

7. Alma Swan, Sheridan Brown, *Authors and Electronic Publishing: The Alpsp Research Study on Authors' and Readers' Views of Electronic Research Communication/Alpsp* (West Sussex, UK: The Association of Learned and Professional Society Publishers, 2002).

8. Barschall, "What's Wrong with These Conferences?" p. 79-81.

9. John Tagler, "204.1 Reply: Newsletter on Serials Pricing Issues, No. 200," *Newsletter on Serials Pricing Issues*, no. 204 (1998). Said this with the intent of providing support for the peer-review of these papers. In other words, not all the papers of a conference pass muster.

10. Dana Roth, "200.4 Elsevier's Conference Proceedings in Journals," *Newsletter on Serials Pricing Issues*, no. 200 (1998), Dana Roth, listserv posting to Reedelscustomers-L maintained at University of Texas at Austin, Dec. 19, 2000, and Bernd-Christoph Kaemper, listserv posting to Reedelscustomers-L maintained at University of Texas at Austin, Dec. 12, 2000, and Dana Roth, "239.1 Librarian's Concern About Content of Journal of Molecular Structure," *Newsletter on Serials Pricing Issues*, no. 239 (1999).

11. Joseph J. Branin, Mary Case, "Reforming Scholarly Publishing in the Science: A Librarian's Perspective," *Notices of the AMS* 45, no. 4 (1998).

12. Austin Barnes, "239.3 Professor Barnes' Response to Librarians," *Newsletter on Serials Pricing Issues*, no. 239 (1999), Jaan Laane, "239.2 Professor Laane's Response to Librarians," *Newsletter on Serials Pricing Issues*, no. 239 (1999), Roth, "239.1 Librarian's Concern About Content of Journal of Molecular Structure."

13. Tagler, "204.1 Reply: Newsletter on Serials Pricing Issues, No. 200."

14. Drummond Rennie, "Fourth International Congress on Peer Review in Biomedical Publication," *Nature* 287, no. 21 (2002).

15. Instructive here is a look at the Morgan Stanley report, *Scientific Publishing: Knowledge is Power* (September 30, 2002) on Reed Elsevier as an investment opportunity.

16. Even Barschall 1992 reports the basic expectation that a published paper from a meeting or conference is the record of what was presented there and then.

17. Allen, "The Magnitude of Conference Proceedings in Physics Journals."

18. Mark S. Frankel, *Seizing the Moment Scientists' Authorship Rights in the Digital Age: Report of a Study by the American Association for the Advancement of Science* (Washington, D.C.: American Association for the Advancement of Science, 2002).

19. Robert K. Merton, "Science and Technology in a Democratic Order," *Journal of Legal and Political Sociology* 1 (1942).

20. Anne M. Buck, Richard C. Flagan, "The Scholar's Forum Debuts," *On the Horizon: A scholarly publication that focuses on the future of postsecondary education* (2002). Provides additional description of the project.

21. Ginsparg's presentation at the SPARC Institutional Repository Workshop in Oct. 2002 included data, verbally presented, that peer-review publishing ranges from $500/paper to even $5,000 per paper. In contrast the XXX preprint server puts the cost per paper at less than $1, including "archiving."

22. The Conference convener made this comment in Sept. 2003.

23. Euan M. Scrimgeour, "Rough Guide to Organising a Medical Congress," *Lancet (North American edition)* 358, no. 9296 (2001).

BIBLIOGRAPHY

Allen, Robert S. "The Magnitude of Conference Proceedings in Physics Journals." *Special Libraries*, no. 86 Spring (1995): 136-44.

Barnes, Austin. "239.3 Professor Barnes' Response to Librarians." *Newsletter on Serials Pricing Issues*, no. 239 (1999). Available from http://www.lib.unc.edu/prices/1999/PRIC239.HTML#239.3.

Barschall, Henry H., Willy Haeberli. "What's Wrong with These Conferences?" *Physics Today*, no. 45 December (1992): 79-80.

Branin, Joseph J., Mary Case. "Reforming Scholarly Publishing in the Sciences: A Librarian's Perspective." *Notices of the AMS* 45, no. 4 (1998): 475-86. Available from http://www.ams.org/notices/199804/branin.pdf.

Buck, Anne M., Richard C. Flagan. "The Scholar's Forum Debuts." *On the Horizon: A scholarly publication that focuses on the future of postsecondary education* (2002). Available from http://resolver.caltech.edu/caltechLIB:2002.003.

Frankel, Mark S. *Seizing the Moment: Scientists' Authorship Rights in the Digital Age: Report of a Study by the American Association for the Advancement of Science.* Washington, D.C.: American Association for the Advancement of Science, 2002. Available from http://www.aaas.org/spp/sfrl/projects/epub/epub.htm.

Kaemper, Bernd-Christoph. Listserv posting to Reedelscustomers-L maintained at University of Texas at Austin, Dec. 12, 2000.

Laane, Jaan. "239.2 Professor Laane's Response to Librarians." *Newsletter on Serials Pricing Issues*, no. 239 (1999). Available from http://www.lib.unc.edu/prices/1999/PRIC239.HTML#239.2.

Malinowsky, H. Robert, Richard A. Gray, Dorothy A. Gray. *Science and Engineering Literature.* 2nd ed. Littleton, CO: Libraries Unlimited, Inc., 1976.

Merton, Robert K. "Science and Technology in a Democratic Order." *Journal of Legal and Political Sociology* 1 (1942): 115-26.

Rennie, Drummond. "Fourth International Congress on Peer Review in Biomedical Publication." *Nature* 287, no. 21 (2002): 2759-60.

Roth, Dana. "200.4 Elsevier's Conference Proceedings in Journals." *Newsletter on Serials Pricing Issues*, no. 200 (1998). Available from http://www.lib.ncsu.edu/stacks/n/nspi/nspi-ns200.txt.

_____. "239.1 Librarian's Concern About Content of Journal of Molecular Structure." *Newsletter on Serials Pricing Issues*, no. 239 (1999). Available from http://www.lib.unc.edu/prices/1999/PRIC239.HTML#239.1.

_____. Listserv posting to Reedelscustomers-L maintained at University of Texas at Austin, Dec. 19, 2000.

_____. "Extending the Online Catalog." In *Questions and Answers: Strategies for Using the Electronic Reference Collection*, 34-37. Urbana-Champaign: Univ. of Illinois, 1987.

Scrimgeour, Euan M. "Rough Guide to Organising a Medical Congress." *Lancet (North American edition)* 358, no. 9296 (2001): 1918. Available from http://www.thelancet.com/.

Swan, Alma, Sheridan Brown. *Authors and Electronic Publishing: The ALPSP Research Study on Authors' and Readers' Views of Electronic Research Communication/ALPSP.* West Sussex, UK: The Association of Learned and Professional Society Publishers, 2002. Available from http://www.alpsp.org/pub5.htm.

Tagler, John. "204.1 Reply: Newsletter on Serials Pricing Issues, No. 200." *Newsletter on Serials Pricing Issues*, no. 204 (1998). Available from http://www.lib.nscu.edu/stacks/n/nspi/nspi-ns204.txt.

Electronic Theses and Dissertations: Enhancing Scholarly Communication and the Graduate Student Experience

Susan Hall

SUMMARY. A growing number of institutions are initiating new programs in electronic theses and dissertations (ETDs). Advances in the ETD as a form of scholarly communication must acknowledge the years of groundwork and planning by representatives from a wide spectrum of organizations such as the Networked Digital Library of Theses and Dissertations (NDLTD), Virginia Tech, the Coalition for Networked Information and the Council of Graduate Schools. More than a decade of multi-institutional collaborative efforts resulted in the current opportunities for students to engage in scholarly communications in new arenas. Where do we go from here? A significant factor in further promoting the ETD and other digital library projects involves shifting program emphasis to key participants, the graduate students who author these documents. *[Article copies available for a fee from The Haworth Document Delivery Service: 1-800-HAWORTH. E-mail address: <docdelivery@haworthpress.com> Website: <http://www.HaworthPress.com> © 2002 by The Haworth Press, Inc. All rights reserved.]*

Susan Hall, BA, MLIS, is Associate Professor, Coordinator, MSU Theses/Dissertations, and Document Delivery Librarian, Mississippi State University.

Address correspondence to: Susan Hall, P.O. Box 5121, MSU, MS 39762 (E-mail: shall@library.msstate.edu).

[Haworth co-indexing entry note]: "Electronic Theses and Dissertations: Enhancing Scholarly Communication and the Graduate Student Experience." Hall, Susan. Co-published simultaneously in *Science & Technology Libraries* (The Haworth Information Press, an imprint of The Haworth Press, Inc.) Vol. 22, No. 3/4, 2002, pp. 51-58; and: *Scholarly Communication in Science and Engineering Research in Higher Education* (ed: Wei Wei) The Haworth Information Press, an imprint of The Haworth Press, Inc., 2002, pp. 51-58. Single or multiple copies of this article are available for a fee from The Haworth Document Delivery Service [1-800-HAWORTH, 9:00 a.m. - 5:00 p.m. (EST). E-mail address: docdelivery@haworthpress.com].

10.1300/J122v22n03_06

KEYWORDS. Electronic theses/dissertations, ETDs, scholarly communication, graduate student education, electronic publishing, digital libraries

INTRODUCTION

Graduate students first engage in scholarly communication in a manner and environment that is somewhat unique compared with that of the seasoned researcher. The habits involving research strategies and information seeking are flexible and subject to trial and new strategies. Students participate in scholarship in close consultation with faculty mentors, and later, a committee chair whose primary concern is new content and scholarly standards. Networks of support are available to the graduate student through work with colleagues, university officials, faculty advisors and committee members. As the final draft of the thesis or dissertation receives institutional approval and is judged a significant contribution to scholarship, the expectation is that the student involvement in scholarly communication will continue. However, this close network of daily direct contact is utilized less later as routines in scholarship are set. Are the years of graduate training a unique opportunity for long term change in scholarly communication? How do library programs respond with support services? The student experience of electronic publishing gained through ETD programs offers an emphasis and opportunity to engage in future scholarship under a new paradigm, serving as a catalyst to change the ways graduate students view the link between electronic publishing, profit and scholarship. In order to promote electronic publishing as an alternative, it is beneficial to examine the graduate student experience as influential in mapping these changes.

ADVANCING SCHOLARLY COMMUNICATION

The ongoing development of the ETD is a natural outgrowth of rapid advances in science and technology fields, a trajectory on the right track. The evolution of digital library projects and the ETD have radically altered scholarly communication and the graduate student experience. Literature review and document production are no longer limited to the paper index, spiral notebooks, index cards, paper drafts and archives. As universities have invested resources in electronic libraries, scholars at their desktop now download entire dissertations that once required a significant time investment (and delay) in identifying, locating, acquiring and keeping track of paper versions acquired through Interlibrary Loan. Both time management skills and some good fortune are necessary for the graduate student expected to review and return several bound dissertations on loan for a short week packed with obligations, minus the required return shipping time.

INSTITUTIONAL COLLABORATION

Throughout the 90s, discussions regarding the ETD at conferences and planning meetings covered such questions as encoding and file format, the cost of electronic storage, archiving processes, indexing beyond the author/title level, and building campus consensus for new programs in electronic publishing. The Networked Digital Library of Theses and Dissertations (NDLTD) is one of many digital library initiatives providing alternatives to commercial publishing by means of more direct involvement by researchers. In *Declaring Independence*, SPARC is described as an "international alliance of over two hundred college and research libraries building a more competitive scholarly communication marketplace to address the high cost of information" (www.arl.org/sparc/DI). NDLTD also holds a related mission, representing the efforts of more than one hundred thirty-five academic institutions worldwide to publish digital libraries, promote access to graduate student documents, and further the education of students in electronic publishing. Since the early discussions of 1987, the particulars of archiving and file format, software development and member assistance have been discussed across university and institutional involvement. The NDLTD website (www.ndltd.org) offers membership listings, information from conference presentations and an overview of key roles and accomplishments.

The ETD project is adaptable, scalable and can be customized for institutional needs. The pilot project at Mississippi State University began with the requirement by the College of Engineering that all engineering students publish ETDs. A year later, during 2001, students from other colleges were invited to publish electronic documents. Concerted effort to enlist wider participation and faculty support is a top priority. One perceived barrier to the project expansion is technology assistance to the graduate student. Despite excellent helpdesk and consultation services provided by Information Technology Services and the availability of an instructional technology/media center within the library, the assumption is often held that technology support would fall to the graduate coordinator. This needs to be addressed to further the project goals.

To date, Mississippi State University has one hundred thirty-eight ETDs on the library website. Ten students have chosen to withhold all access to the ETD document. Two students stated concern over proprietary information or patents pending. However, exit surveys were not in place to collect feedback on the other students' decisions.

The rate of growth of Mississippi State's ETD library is significant. During the first full semester of the ETD project, Spring 2001, fewer than twenty documents were published electronically. In Fall 2002, forty-one ETDs were added to the MSU website which includes publications from Veterinary Medicine, Physics, Geosciences, Forest Products and Poultry Science.

GRADUATE STUDENT CONCERNS

There remains some concern regarding the status of the ETD as a published document. A student in chemistry may produce several chapters of the dissertation to be submitted for journal publication. A number of graduate faculty hold the view that publishers may consider the ETD a prior publication. Advisors in the humanities fields may counsel students that book contracts require extended periods of time, and that releasing the dissertation for web publication could undermine long term goals for reworking the dissertation as a book contract. These issues have great significance for the graduate student's academic career.

Wide distribution is often interpreted as inviting plagiarism. However, it may be that such distribution actually provides the antidote. It seems more likely that a single bound copy of a dissertation found in one library archive could be targeted for illegitimate scholarship, and not the document that is widely read and available in an open digital library. Archiving the ETD rather than the paper version offers clear advantages, in that the binding and shelf space costs are erased. However, producing the ETD does involve significant investments in technology, instruction to new students, and designing up-to-date and effective web page information. It is especially important to the graduate student working against difficult time pressures that clear procedures regarding electronic publishing and associated deadlines be made explicit. Frequently mentioned concerns are issues related to the ready availability of technology support and the perception that additional costly upgrades will be necessary. Some of these perceptions can be addressed through frequent meetings with the graduate student association.

BALANCING ROLES

Libraries involved in ETD projects find they engage in a new balancing act. The institutional obligation as a depository for scholarship is balanced with an obligation to pursue active roles in furthering electronic publishing, in some cases providing primary service contact for the student publishing the ETD. This requires institutional commitment beyond traditional unit boundaries. Graduate schools, libraries, and student advisors will need a period of reorganization and flexibility if the project is to succeed. Some institutions still require both formats, providing electronic versions and archiving an unbound paper copy. A number of institutions are embarking upon projects for systematic retrospective conversion. As older dissertation titles that pre-date the ETD project are requested, the distribution process includes a step to make the document available as an ETD. Building comprehensive digital libraries with retrospective collections greatly enhances scholarly communication for the graduate student beginning the literature review, as well as the seasoned researcher.

THE FUTURE OF SCHOLARLY COMMUNICATION

The university setting influences the graduate student's approach to engaging in scholarly communication, offering electronic as well as traditional print resources as tools for conducting research. As training and assistance in electronic publishing is made available to new graduate students, a window to long-term change is provided. The future faculty member provided with technology support that is responsive and readily available enhances the likelihood for future involvement in electronic publishing endeavors. Consultation with an information technology professional is a routine that graduate students can easily establish. It is not as frequently the case that long-tenured faculty across all disciplines would do so. As numbers of non-traditional and distance education students increase, the 24/7 access to digital libraries is an important feature for those juggling obligations to family, career and scholarship.

CONTINUED PROGRESS IN ETDs

With a recent review of university web page information regarding ETDs, it is clear that progress is being made in this form of scholarly communication. To date, more than twenty-five U.S. institutions have established ETD programs providing web pages with details regarding publication requirements, support, and in most cases, a significant library of theses and dissertations available in full text. These web presences may evolve slowly with sites under construction for extended periods, or they may be unveiled overnight, fully developed with complete information, FAQs, and submission instructions to students. It is impossible to envision this rapid change without acknowledging the careful discussion and planning of key collaborators such as Virginia Tech, the Networked Digital Library of Theses and Dissertations, Southeastern Universities Research Associations, UMI/ProQuest, and OCLC:

> The first workshop about electronic theses and dissertations (ETDs) took place in 1987 with a technical focus on standards, namely applying SGML to the description of research. Ten years later, we realize that the proper aim should be improving graduate education by having students enter ETDs into a digital library which facilitates much broader access. Achieving that goal calls for a sustainable, worldwide, collaborative, educational initiative of universities committed to encouraging students to prepare electronic documents and to use digital libraries–NDLTD. (Fox, Eaton, McMillan, Kipp et al. 1997)

After discussions that centered on technology, institutional change, planning for new responsibilities, and issues regarding copyright and prior publications, the focus shifts to building services for the authors of ETDs.

CONCLUSION

How does the graduate student first engage in scholarly communication? Throughout the years of study, an ongoing and systematic approach to review of the literature is expected. An initial concern is to clear a proposal with committee members that shows promise for originality. Collection and assessment of data may generate refinements to the proposal. Completing the research project before another scholar addresses the topic becomes the next immediate concern. The literature review utilizing comprehensive databases to identify a unique niche for the proposal or dissertation is assumed. How does the student make use of electronic resources? Often consultations and comparisons of results are shared with faculty, committee members, colleagues and librarians to ensure complete coverage of cited literature. Just as fully developed reference services are provided to the student conducting a literature review in a specialized database, a library service component addressing electronic publishing is also appropriate. In many cases graduate students spend hours or days trying to force software to comply with university guidelines for thesis/dissertation format. Expert staff often see the same questions each semester. By providing and promoting technology assistance, a great deal of the time spent on bad fits between traditional presentation styles and software quirks can be reduced. Instruction services that are effective during this last phase of campus experience can greatly enhance the overall graduate student experience. Finding the appropriate institutional niche for supporting student efforts in electronic publication will be a long-range investment with a high impact on scholarly communication.

REFERENCES

Erickson, Janet. "An SGML/HTML Electronic Thesis and Dissertation Library." *Text Encoding Initiative, 10th Anniversary User Conference.* (1997): available at: www.stg.brown.edu/conferences/tei10/tei10.papers/erickson.html (accessed March 7, 2003).

Fox, Edward A., John L. Eaton, Gail McMillan, Neill A. Kipp, Paul Mather, Tim McGonigle, William Schweiker, Brian DeVane "Networked Digital Library of Theses and Dissertations: An International Effort Unlocking University Resources." *D-Lib Magazine.* (1997) available at: www.dlib.org/dlib/september97/theses/09fox.html (accessed March 7, 2003).

Fox, Edward A. "Improving Graduate Education with the National Digital Library of Theses and Dissertations." (27 June 1992): available at: http://www.ndltd.org/pubs/FIPSEfr.pdf (accessed March 7, 2003).

Fox, Edward A., Gail McMillan, John L. Eaton. "The Evolving Genre of Electronic Theses and Dissertations." *32nd Annual Hawaii International Conference on Systems Sciences (HICSS) Maui, HI–January 5-8, 1999.* (1999): available at www.ndltd.org/pubs/Genre.htm (accessed March 7, 2003).

McMillan, Gail. Librarians as Publishers: Is the Digital Library an Electronic Publisher? *College & Research Library News.* 61, no. 10 (2000): p. 928-931.

McMillan, Gail. "Managing Electronic Theses and Dissertations: The Third International Symposium." *College & Research Library News.* 61, no. 5 (2000): p. 413-414.

McMillan, Gail. "Do ETDs Deter Publishers? Coverage from the 4th International Symposium on ETDs." *College & Research Libraries News.* 62, no. 6 (2001): p. 620-621.

McMillan, Gail. "Librarians as Publishers: Is the Digital Library an Electronic Publisher?" *College & Research Library News.* 61, no. 10 (2000): 928-931.

Newby, Jill. "STS General Discussion Group: Summary of ALA Midwinter Meeting, January 10, 1998." *Issues in Science & Technology Librarianship* 17 (Winter

1998): available at http://www.library.ucsb.edu/istl/98-winter/conference3.html (accessed March 7, 2003).

Sharretts, Christina W., Jackie Shieh, James C. French. "Electronic Theses and Dissertations at the University of Virginia." *Proceedings of the ASIS Annual Meeting.* 36 (1999): p. 240-255.

Suleman, Hussein, Anthony Atkins, Marcos A. Goncalves, Robert K. France, Edward A. Fox, Vinod Chachra, Murray Crowder, and Jeff Young. "Networked Digital Library of Theses and Dissertations. Bridging the Gaps for Global Access–Part 1: Mission and Progress." *D-Lib Magazine.* (2001): available at: www.dlib.org/dlib/september01/suleman/09suleman-pt1.html (accessed March 7, 2003).

Websites:

http://www.ndltd.org
http://www.theses.org
http://www.arl.org/sparc/DI
http://www.cni.org.

Chemistry Journals:
Cost-Effectiveness, Seminal Titles
and Exchange Rate Profiteering

Dana L. Roth

SUMMARY. The cost-effectiveness of STM journals has been compared within several subject areas, beginning with Henry Barschall's work with the physics literature in the late 1980s. A new use-independent cost-effectiveness metric is proposed and calculated for journals in several chemistry subdisciplines.

Publisher and year-of-publication data for seminal journal articles assigned in a graduate-level organic synthesis class are presented.

The effects of publisher policies in establishing and enforcing differential subscription prices for European and non-European customers on the rise of

Dana L. Roth, BS, MS, MLS, is a member of the Professional Staff, Senior Technical Information Librarian, and Chemistry Librarian, Caltech (E-mail: dzrlib@library. caltech.edu).

Address correspondence to Dana L. Roth, 1200 East California Boulevard, Pasadena, CA 91125.

Ken Frazier and Ken Rouse's work at the University of Wisconsin-Madison provided the inspiration for thinking about cost-effectiveness in a new way. Bernd-Christoph Kaemper at the Universitätsbibliothek Stuttgart was a source of excellent advice and additional references. Robert Michaelson at Northwestern University assisted with data collection.

[Haworth co-indexing entry note]: "Chemistry Journals: Cost-Effectiveness, Seminal Titles and Exchange Rate Profiteering." Roth, Dana L. Co-published simultaneously in *Science & Technology Libraries* (The Haworth Information Press, an imprint of The Haworth Press, Inc.) Vol. 22, No. 3/4, 2002, pp. 59-70; and: *Scholarly Communication in Science and Engineering Research in Higher Education* (ed: Wei Wei) The Haworth Information Press, an imprint of The Haworth Press, Inc., 2002, pp. 59-70. Single or multiple copies of this article are available for a fee from The Haworth Document Delivery Service [1-800-HAWORTH, 9:00 a.m. - 5:00 p.m. (EST). E-mail address: docdelivery@haworthpress.com].

10.1300/J122v22n03_07

journal subscription costs and also on possible exchange-rate profiteering are discussed. *[Article copies available for a fee from The Haworth Document Delivery Service: 1-800-HAWORTH. E-mail address: <docdelivery@haworthpress.com> Website: <http://www.HaworthPress.com> © 2002 by The Haworth Press, Inc. All rights reserved.]*

KEYWORDS. Cost-effectiveness, chemistry, journals, foreign exchange rates, organic synthesis

COST-EFFECTIVENESS OF SOME CHEMISTRY JOURNALS

Significant author subsidization of the publication of chemistry and physics journals, through page charges, has steadily eroded over the past forty years, largely because of no-charge policies of commercial publishers. Both the American Chemical Society (who also do not charge authors for cover art or 4 color process) and American Physical Society have recently dropped page charges, transitioning completely to the "reader pays" model. This has the unintended effect of allowing society journal subscription rates to be directly compared with those of commercially published journals, and invites a comparison of their relative cost-effectiveness.

Henry Barschall's pioneering study of the cost-effectiveness of physics journals (1) prompted studies in a variety of other subject areas, especially chemistry (2). However, analyses of the cost-effectiveness of journals in very narrowly defined subject areas, and independent of usage, has received very little attention. A use-independent approach contradicts the assertion that "all researchers have agreed that a print journal's value cannot be assessed with content evaluation alone" (2e). However, once a user group has identified its core journals, development of a use-independent metric should prove to be useful for evaluating the cost-effectiveness of these titles.

In 1995, the California Institute of Technology undertook a significant review of its journal holdings. Professorial faculty identified those journals essential to their research, and a funding mechanism was established to ensure their continued availability. The use of these titles is now systematically reviewed, resulting in a few cancellations. Newly published titles are regularly added, again with the understanding that they are essential for a specific faculty member's research projects. Established titles are added if the interlibrary loan rate warrants subscription.

Beginning in 2000, with the transition from exchange-rate subscription pricing to unlinked US$, Yen, and EURO pricing, for commercially published STM journals, it became obvious that annual 5-10% subscription price increases will not be sustainable. Since a 10% annual price increase equates to a doubling time in seven years, a new rigorous methodology to evaluate commercially published journals was required.

A cost-effectiveness metric that can be used to compare journals that are publishing equivalent content should be of significant benefit to both librarians and library users. This is particularly true since the advent of large journal packages (e.g., *Science Direct*) and the recent recognition that there is a "true (STM publishing) market failure" (3). Since user demand for electronic access is rarely based on either price or quality considerations, a use-independent metric should prove to be a very effective tool for assessing and prioritizing user demands. As an aside, it should be obvious that each established journal enjoys a monopolistic position vis-à-vis its own subscribers and readers, and that the reverse is true with respect to its own authors and editorial boards. This dichotomy is illustrated, on the one hand, by libraries (subscribers) that continue to purchase obviously overpriced titles (4) at the insistence of their users (readers), and on the other hand by the dramatic resignations of editorial boards from established journals (5a), with the intent to re-establish a cost-effective product.

The current extraordinarily high cost of commercially published STM journals (5b) is based on a wide variety of factors, but is not justifiable when compared with society published STM journals. Publishers do not fund research, do not provide monetary advances to authors, and are not responsible for the editorial selection of journal articles. Rapidly increasing automation of both the submission of articles and the mechanics of referring (6) suggests that the publisher's role will soon be (or should be) simplified to the role of providing minimal copy editing, maintaining an electronic infrastructure, and hiring account managers to process subscriptions and advertisements. This should then eliminate disparities in both subscription prices and cost-effectiveness of journals that publish equivalent information. Continuing disparities in price and cost-effectiveness must then be carefully examined and brought to the attention of librarians, readers, authors and editors.

The importance of comparing journals publishing equivalent information, when discussing costs and quality with faculty members or administrative staff, cannot be underestimated. Previous studies on journal costs and effectiveness focused on fairly general subject areas and did not highlight the differences between journals that publish equivalent information of interest to individual research groups.

In an attempt to formulate a clearer, more illustrative metric for determining cost-effectiveness, the ratio of normalized values for "cost" and "impact" are calculated for chemistry journals publishing equivalent material in 2001. Normalization is simply the rescaling of a set of values, e.g., dividing each journal's ISI Impact Factor by the Impact Factor of a baseline journal. This results in the baseline journal having an relative Impact Factor of unity and quickly provides rescaled and more easily understandable comparative values, for other journals, relative to the baseline journal.

Values of the new "Cost-Effectiveness" metric are shown in Table 1. They are analogous to Barschall's (1), Christensen's (2c), and Rouse's (2d), but differ in that each journal's cost-per-page (7) and ISI Impact Factor (8) are both normalized to a baseline journal. Additional comparisons can then be made by simply renormalizing to a different baseline. It is also interesting to note (in Table 1) the generally strong inverse relationship between subscription cost per page and ISI Impact Factor.

Assuming that the ratio of the normalized Cost/Page to the normalized Impact Factor (NCP/NIP) is an effective cost-effectiveness metric, one could conclude, for example, that (in 2001) inorganic information in the ACS' *Inorganic Chemistry* was over fifteen times as cost-effective as "inorganic" information packaged in *Inorganica Chimica Acta*. Similarly, "chemical physics" information packaged in the AIP's *Journal of Chemical Physics* was eleven times more cost-effective than that in *Chemical Physics*, and "physical organic" information packaged in *Journal of Chemical Society Perkin II* was four times more cost-effective than that in *Journal of Physical Organic Chemistry*.

Additional comparisons can be made by renormalizing to a different baseline journal. For example, "inorganic" information packaged in *Journal of the Chemical Society Dalton Transactions* was nearly six times as cost-effective as that in *Polyhedron*. Similarly, "chemical physics" information in *Physical Chemistry Chemical Physics* was three times as cost-effective as that in *Chemical Physics*.

Assuming *Inorganic Chemistry* is 15.45 times as cost-effective as *Inorganica Chimica Acta,* suggests that to be as cost-effective as *Inorganic Chemistry*, *Inorganica Chimica Acta*'s 2001 subscription cost would have been $436 (instead of $6,726).

These examples are surprising in that the print subscription price paid by large academic libraries for *Inorganic Chemistry*, *Journal of Chemical Physics*, *Physical Chemistry Chemical Physics* and *Journal of Chemical Society Perkin II* all include multiyear site-wide electronic access, whereas subscriptions to *Polyhedron* and *Chemical Physics* include only

TABLE 1. 2001 Cost per Page, Impact Factor, and Cost-Effectiveness Data for Chemistry Journals

JOURNAL	C/P	NC/P	IP	NIP	NCP/NIP
Inorganic Chemistry	$0.27	1.00	2.95	1.00	1.00
J. Chem. Soc. Dalton	$0.64	2.37	2.82	0.96	2.47
Eur. J. Inorg. Chem.	$0.65	2.41	2.48	0.84	2.87
Polyhedron	$1.60	5.93	1.20	0.41	14.46
Inorganica Chim. Acta	$1.96	7.26	1.39	0.47	15.45

J. Chemical Physics	$0.20	1.00	3.15	1.00	1.00
Phys. Chem. Chem. Phys.	$0.40	2.02	1.79	0.57	3.54
Chemical Physics	$1.41	7.08	1.96	0.62	11.41

J. Chem. Soc. Perkin II	$0.88	1.00	1.84	1.00	1.00
J. Physical Org. Chem.	$1.75	2.64	1.30	0.71	3.72

C/P = Cost/Page, NC/P = Normalized C/P, IP = 2001 ISI Impact Factor, NIP = Normalized IP, NCP/NIP = Normalized Cost/Page/Normalized Impact Factor = Cost-Effectiveness

a one-year, rolling-window, site-wide electronic access, and *Journal of Physical Organic Chemistry* provides, as one option, a one-seat multiyear electronic access at a 5% premium.

In this way, the dramatic differences–in page costs, impact factors, cost-effectiveness and electronic access–between society journals and those of some commercial publishers can be easily understood. Generally, societies operate as non-profit entities (albeit with any journal subscription profit going to support other society activities), while some large commercial publishers, other than their attempts to balance subscription costs with cancellations, seemingly have no restraint on their profitability (9).

DATA FOR SEMINAL ORGANIC SYNTHESIS ARTICLES

In addition to the development of a new cost-effectiveness metric, it is also interesting to compare the distribution of references, title by title, to seminal organic synthesis articles. The comparison, in Table 2, is based on the 225 journal articles assigned in Caltech's (Fall 2002) graduate-level Chem 242–Chemical Synthesis class. It is interesting to note that over 72% of the articles were published in society or society-sponsored journals, and only 24% were published in commercial journals.

This implied quality assessment for ACS journals is further substantiated by the fact that 4 of the 5 most highly cited articles reporting new

TABLE 2. Source Journals for Assigned Readings in Chem 242–Chemical Synthesis (225 Articles)

J. Am. Chem. Soc. (98)	(43.5%)
J. Org. Chem. (25), Org. Lett. (3)	(12.5%)
Angew. Chem. Int. Ed. (19)	(8.4%)
Chem. Rev. (9), Accts. Chem. Res. (3)	(5.3%)
J. Chem. Soc. (2), Chem. Commun. (4)	(2.7%)
Tet. Lett. (29), Tetrahedron (7), Tet. Asymmetry (1)	(12%)
Org. Rxn. (8), Org. Syn. (2)	(4.4%)
Synthesis (6), SynLett (1)	(3.1%)
Eight miscellaneous titles (1 each)	(3.5%)

organic synthetic techniques from 1985 thru 1996 came from JACS and the 5th was from JOC (10).

It is also interesting to note the increasing number of articles published in society-sponsored journals by five-year segments since 1982. Table 3 suggests that the frequency of publication of seminal articles in non-society related journals is decreasing.

US$ SUBSCRIPTION PRICES AND FOREIGN CURRENCY EXCHANGE RATES

One largely misunderstood factor in the cost-per-page disparities (shown in Table 1) has been the effect of foreign-currency fluctuations, beginning in the early 1970s. Since many commercial publishers of scientific journals are based in Europe, fluctuations in the U.S. dollar cost of continental European currencies, combined with what appears to be exchange-rate profiteering, must be clearly understood. Figure 1 gives the year-to-year average cost of Dutch guilders (NLG) since 1970 (11).

In 1970, a Dutch guilder (NLG) could be purchased for $0.276 and, for example, a 1000-NLG subscription would have been billed at US$276. Contrast this with the years from 1971 to 1980, when the cost of a guilder steadily increased to about $0.503 and, with all other factors being equal, a 1000-NLG subscription would have increased to $503. This was the first "serials crisis," and generally resulted in libraries canceling duplicate subscriptions, reducing book purchases, and effecting other cost-containment strategies. In contrast, however, from 1981 to

TABLE 3. Distribution of Publication Dates for Selected Titles (1983-2002)

	1983-87	1988-92	1993-97	1998-2002
ACIEE/ACIE	4	2	1	11
ACS	19	24	17	15
Tet./Tet. Lett.	11	6	1	4

1985, the cost of a guilder steadily decreased to about $0.301, and this should have resulted in steadily decreasing subscription rates. This rise and fall of exchange rates is a widely understood phenomena and with some institutional investment flexibility can be accommodated. During this 1981-1985 period, when the purchasing power of the U.S. dollar was dramatically increasing, it obviously would have been a substantial benefit to U.S. subscribers, ignoring inflation, to have the NLG1000 base rate remain in effect through 1985, as this would have resulted in lowering the subscription rate to US$301. However, as the US$ cost of the guilder dropped in the years 1981 to 1985, guilder subscription rates were increased to maintain a relatively constant US$ subscription rate. The net effect was to artificially lock in an additional ~67% to the base subscription rate. This increase became a continuing annual cost, and had a significant multiplier effect when the U.S. dollar cost of the guilder began increasing again in mid-1985.

By year end 1986, the U.S. dollar cost of the guilder had had a one-year increase of ~30% (from $0.301 to $0.408); this was only the beginning of a second dramatic rise in the US$ cost of a guilder, which by 1995 had increased to $0.624. This second "serials crisis" was devastating for two reasons: the significant increase in base subscription rates during the 1981-1985 period, and the confusing difference between the cost of the guilder and the value of the U.S. dollar.

Exchange rate quotations are generally stated in the popular press in terms of the number of guilders one can purchase for a U.S. dollar, rather than the more useful (for librarians), U.S. dollar cost of a guilder. The confusion is best exemplified by comparing a change from 3.2 guilders per U.S. dollar to 1.6 guilders per U.S. dollar. This is typically described as being a 50% decrease in the value of the U.S. dollar, but in actual fact is a 100% increase in the U.S. dollar cost of a guilder ($0.3125 vs. $0.625).

During 1996, the guilder again began declining (as one would expect from the cyclic nature of foreign exchange rates), but, presumably because of the already excessive guilder subscription rates, publishers could not substantially raise guilder rates to maintain a relatively con-

FIGURE 1. U.S. Dollar Cost of Dutch Guilders, 1970-2002

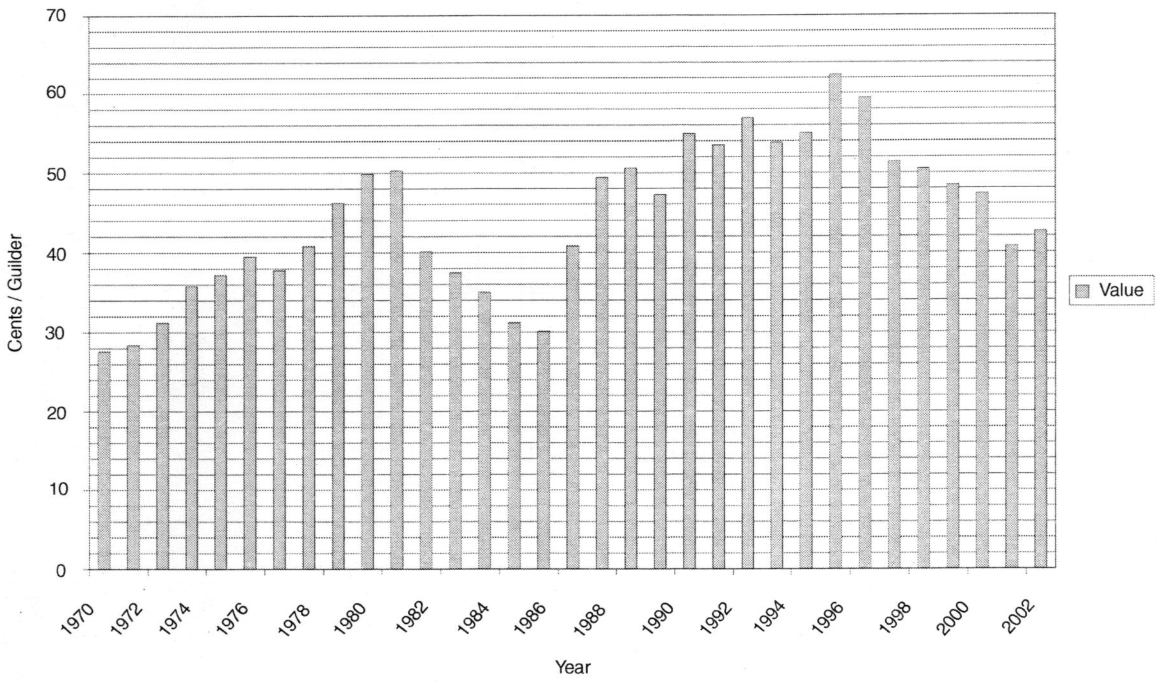

stant US$ rate. Then, in 1999, most European commercial publishers adopted a new tactic for the 2000 subscription year: This time, instead of raising the NLG price to reflect the U.S. dollar price, they adopted a new policy that invoices non-European customers in US$ (and Japanese customers in YEN), using the 1999 US$ price as a base and no longer allowing subscriptions to be placed with European agents at the "exchange rate NLG" price, as had previously been possible.

This policy was a complete reversal from previous years (12). U.S. library subscribers, instead of benefiting from continuing guilder exchange-rate declines in 2000 and 2001, were only promised that the U.S. dollar subscription rates would increase by less than 10%. These new "Elsevier" U.S. dollar subscription rates appear to be completely disconnected from the NLG/EURO rates (13), and deny U.S. dollar subscribers the savings that would have resulted under previous exchange-rate-based pricing policies. The net effect resulted in what one might describe as substantial exchange-rate profits. The new "Millennium" subscription rate policy is shown in Table 4.

In conclusion, the cost-effectiveness analysis corroborates both Bensman's major finding (14), "that little relationship . . . exists between scientific value and the prices charged libraries for scientific journals," and Hahn and Faulkner's observation (2e) that "What a publisher charges for a particular journal does not necessarily reveal anything about its relative value." In addition, it should be obvious that the escalating price increases of commercially published journals cannot be sustained. This is especially true when non-European subscribers suffer the decline in value of the U.S. dollar, but are subsequently denied the benefits of its appreciation. Given the excellent cost-effectiveness of society and society-sponsored journals, these are obviously the titles that must be retained, and less cost-effective titles must be identified and subjected to further evaluation.

TABLE 4. Exchange Rate Profits for *Tetrahedron*, 2000-2003

Tetrahedron (including: Tetrahedron Asymmetry):

Year	NLG price	Exch rate / US$ price	'ELS' US$ price	Exch Rate Profit	(%)
2000	NLG 22899	$0.50 / $11449	$11624	$175	(1.5%)
2001	NLG 24440	$0.455 / $11120	$12406	$1286	(11.56%)
2002	NLG 26030	$0.406 / $10568	$13212	$2644	(25.0%)
2003	NLG 27983	$0.407 / $11389	$14203	$2814	(24.7%)

Please note that although the NLG has been replaced by the EURO, a fixed NLG-to-EURO exchange rate of 2.2037 is still in effect.

REFERENCES

1a. H.H. Barschall, "The Cost of Physics Journals," *Physics Today* 39 (12): 34-36, December 1986. http://barschall.stanford.edu/ (accessed 2/15/03).

1b. H.H. Barschall, "The Cost-Effectiveness of Physics Journals," *Physics Today* 41 (7):56-59, July 1988. http://barschall.stanford.edu/ (accessed 2/15/03).

1c. H.H. Barschall, "Cost-Effectiveness of Physics Journals," *Physics Today* 42 (3):15-16, March 1989. http://barschall.stanford.edu/ (accessed 2/15/03).

2a. M.M. Case, "Measuring the Cost Effectiveness of Journals: The Wisconsin Experience," http://www.arl.org/newsltr/205/wisconsin.html (accessed 2/6/03).

2b. G. Soete, "Measuring the Cost-Effectiveness of Journals: Ten Years After Barschall," http://www.library.wisc.edu/projects/glsdo/cost.html (accessed 2/6/03).

This reference includes links to "The Article," as well as data tables for physics, economics and neuroscience. Selected Sources (for other studies) are listed at the end of "The Article."

2c. J.G. Christensen, "Chemistry Journal Costs at One University," *Serials Review* 18 (3): 19-34, Fall 2000.

2d. K. Rouse, "Chemistry Journal Cost Study," http://www.library.wisc.edu/libraries/ Chemistry/cost.htm (accessed 2/6/03).

2e. K.L. Hahn and L.A. Faulkner, "Evaluative Usage-based Metrics for the Selection of E-journals," *College and Research Libraries* 63 (3):215-227, May 2002.

3a. R. Poynder, "A True Market Failure: Professor Mark McCabe talks about problems in the STM publishing industry," *Information Today* 19 (11):1, 56, 58, December 2002. http://www.infotoday.com/it/dec02/poynder.htm (accessed 2/6/03).

3b. M. Cornet and B. Vollard, "Tackling the Journal Crisis," *CPB Working Paper* 121, March 2000. http://www.cpb.nl/eng/pub/werkdoc/121 (CPB Netherlands Bureau for Economic Policy Analysis, The Hague, The Netherlands) (accessed 2/6/03).

4a. D. Bradley, "Journal Publishers to Police Themselves," *Information Today* 16 (21) October 28, 2002, http://www.the-scientist.com/yr2002/oct/prof2_021028.html (accessed 2/6/03).

4b. Reed-Elsevier Financial Highlights (for the six months ending June 30, 2001) http://www.reedelsevier.com/investors/2001/interim2001.pdf (accessed 2/6/03).

4c. R. Morais, "Double Dutch No longer; Amid a media recession, Crispin Davis is coining money at Reed Elsevier. How did he pull that off?" *Forbes.com* (11/11/01) http://www.forbes.com/global/2002/1111/044.html (accessed 2/6/03).

4d. P. Gooden et al., "Scientific Publishing: Knowledge is Power" Morgan Stanley Industry Overview (Sept. 30, 2002). Investext Database (accessed 2/6/03).

5a. A. Buckholtz, "Returning Scientific Publishing to Scientists," *Journal of Electronic Publishing* 7, no.1 (August 2001*).* http://www.press.umich.edu/jep/07-01/buckholtz. html (accessed 2/6/03).

5b. A notable exception is Thieme Medical Publishers, Inc. publishers of SynLett and Synthesis. See: K. Kurz, "Pricing of Chemistry Journals–A Comparison of Journals Published by Not-for Profit Organizations and Commercial Publishers." *Newsletter on Serials Pricing Issues* 249 (June 16, 2000). 249.5. http://www-mathdoc. ujf-grenoble.fr/NSPI/Numeros/2000-249.html#5.

6. "ACS Paragon System; new journal manuscript submission system." *Chemical & Engineering News*. 80 (49):9, December 9, 2002.

7. Because of the wide variety of pricing models, the print subscription cost of each journal was used. This has the disadvantage of penalizing the Royal Society of Chemistry and the American Institute of Physics, since they include electronic access at no additional charge. This is in marked contrast with commercial publishers, who require additional payment for comparable electronic access. Societies are at an additional disadvantage because they also provide package discounts. Further analysis to compensate for these differences is required. In addition, pages were assumed to be equivalent although the print density and page size of commercial journals is often less than that of society journals.

8a. "The Impact Factor." Institute for Scientific Information. http://www.isinet.com/isi/hot/essays/journalcitationreports/7.html (accessed (2/6/03).

8b. "M. Amin and M. Mabe, "Impact Factors: Use and Abuse." *Perspectives in Publishing* no.1 (October 2000). http://www.elsevier.com/homepage/about/ita/editors/perspectives1.pdf (accessed 2/12/03). Amin and Mabe's main point is that "the value of the impact factor is affected by factors that include the subject area of the journal, the type of journal (letters, full papers, reviews) . . ." This study was designed to avoid these problems.

9a. Robert Maxwell (former owner of Pergamon Press, who died in 1991) is reported to have said, "that even if all libraries, save one, cancelled their subscriptions to his journals, he could still make money by raising the price of that last subscription very, very high." I.E. McDermott, "Confessions of a Serial Clicker." *Searcher* 10 (9), 8, October 2002.

9b. E. Garfield, "Why Scientific Journals should be Audited," *Essays of an Information Scientist* 14:354-355, 1991. http://www.garfield.library.upenn.edu/essays.html (accessed 2/6/03).

10. M. Clark, "Synthetic chemical literature." *J. Chem. Inf. Comput. Sci.* 1999, 39, 635-637. http://pubs.acs.org/cgi-bin/article.cgi/jcisd8/1999/39/i04/pdf/ci990001e.pdf.

11a. *International Financial Statistics Yearbook*, 1995. Washington, D.C., International Monetary Fund, c1995.

11b. Oanda.com (The Currency Site). http://www.oanda.com/convert/fxhistory (accessed 12/11/02).

12a. D. L. Roth, "US Dollar vs Dutch Guilder." *Newsletter on Serials Pricing Issues* 200 (January 29, 1998). 200.2. http://www.lib.unc.edu/prices/1998/PRIC200.HTML#200.2 (accessed 2/6/03).

12b. J. Tagler, "Reply: Newsletter on Serial Pricing Issues, No. 200." *Newsletter on Serials Pricing Issues 204* (March 13, 1998). 204.1. http://www-mathdoc.ujf-grenoble.fr/NSPI/Numeros/1998-204.html (accessed 2/6/03).

13. D. Haank, "New Elsevier Science Journal Pricing Policy to Accompany Transition from Print to Digital Delivery," *Newsletter on Serial Pricing Issues*, no. 227 (June 14, 1999). 227.2. http://www.lib.unc.edu/prices/1999/PRIC227.HTML (accessed 2/6/03).

14. S.J. Bensman, "The Structure of the Library Market for Scientific Journals: The Case of Chemistry." *Library Resources & Technical Services* 40 (2):145-170, April 1998.

ADDITIONAL RESOURCES

P. Brueggeman, "Impact of Scientific Journal Costs," http://scilib.ucsd.edu/sio/guide/prices/ (accessed 2/6/03).

SPARC, "Declaring Independence," Appendix B. http://www.arl.org/sparc/DI/appendixB.html (accessed 2/6/03).

D.L. Roth, "Pricing of Chemistry Journals," *Newsletter on Serials Pricing Issues*, no. 253, 9/22/2000. http://www-mathdoc.ujf-grenoble.fr/NSPI/Numeros/2000-253.html (accessed 2/6/03).

D.L. Roth, "Differential Pricing and Exchange Rate Profits," *IATUL Proceedings* (New Series) Vol. 12, 2002. http://www.iatul.org/conference/02/Roth.pdf (accessed 2/6/03).

The Role of Scientific Literature in Electronic Scholarly Communication

John Cruickshank

SUMMARY. The perception that advanced information technology necessarily improves the effectiveness of scientific and scholarly research overlooks serious weaknesses in the scholarly communication circuit of scientists. By examining models of the scholarly communication circuit of scientists developed in communication science, and models of information flow developed in the fields of library and information science, bottlenecks in the flow of information between scholars can be identified and possible solutions to underlying problems can be considered.

Significant potential bottlenecks to the flow of information become especially evident upon close examination of certain key roles of the scientific literature in the scholarly work of scientists. By considering the

John Cruickshank, BSc, MScF, MLIS, is Interim Librarian, College of Veterinary Medicine Library, Mississippi State University, P.O. Box 6204, Mississippi State, MS 39762 (E-mail: jlcruickshank33@msn.com).

[Haworth co-indexing entry note]: "The Role of Scientific Literature in Electronic Scholarly Communication." Cruickshank, John. Co-published simultaneously in *Science & Technology Libraries* (The Haworth Information Press, an imprint of The Haworth Press, Inc.) Vol. 22, No. 3/4, 2002, pp. 71-100; and: *Scholarly Communication in Science and Engineering Research in Higher Education* (ed: Wei Wei) The Haworth Information Press, an imprint of The Haworth Press, Inc., 2002, pp. 71-100. Single or multiple copies of this article are available for a fee from The Haworth Document Delivery Service [1-800-HAWORTH, 9:00 a.m. - 5:00 p.m. (EST). E-mail address: docdelivery@haworthpress.com].

essential types of information that scientific literature is meant to communicate, and by examining strengths and weaknesses of the scholarly communication circuit in communicating that information, information specialists can consider alternative and more effective approaches to meeting the scholarly information needs of scientists.

Examination of such issues as the role of scientific documents in social processes of scientific scholarship, the problems associated with how meaning is represented by documents, the problems associated with the fragmentation of scientific literature as a result of the growth of literature, and the difficulties associated with the use of information retrieval systems, highlight the serious limitations of the scholarly communication circuit, irrespective of the use of advanced information technology.

Implications of these issues for the librarian and the scientific scholar are discussed. *[Article copies available for a fee from The Haworth Document Delivery Service: 1-800-HAWORTH. E-mail address: <docdelivery@haworthpress. com> Website: <http://www.HaworthPress.com> © 2002 by The Haworth Press, Inc. All rights reserved.]*

KEYWORDS. Scholarly communication, scientific literature, scientific communication, digital libraries, information retrieval, information-seeking behavior

INTRODUCTION

The development of efficient computers and computer networks has received considerable attention in the media in recent years. Such initiatives as Internet II and the Digital Library Initiative show ample evidence of the importance which the scientific community at large attaches to the rapid transfer of scientific information through electronic networks. It is quite clear that the rate at which scientific information can be transferred will continue to increase for some time to come as high-speed fiberoptic networks continue to be installed throughout the world and the microchip increases in power. How much real difference do these advances in information technology make in terms of the quality or quantity of research produced by the individual scientist or scientific communities? Does it logically follow that because greater quantities of information are being transferred through networks at greater rates among scientists and institutions that scientists are necessarily more effective or productive in performing research? Is it possible that the reliance on information technology could in some important ways hinder the scholarly communication of scientists?

In seeking to understand the processes involved in the scholarly communication of scientists, social scientists in the 1960s and 1970s built models of information flows by testing theories of scientists' behavior (Garvey, 1972; Garvey et al., 1979). They established a basic model of the scholarly communication circuit of scientists which, despite recent developments in information technology, is as valid today as it was over three decades ago (Crawford et al., 1996). Because of its complex nature, the scientific literature has turned out to be considerably more problematic as a component of this communication circuit than either scientists or information specialists are generally aware.

By examining common models of the scholarly communication circuit of scientists, and models of the structure and function of scientific literature as a component of that circuit, bottlenecks in the flow of information can be identified and possible solutions to underlying problems can be considered.

COMMON MODELS OF SCHOLARLY COMMUNICATION

Models from Communication Science

In the model of scholarly scientific communication established by Garvey and Griffith (1972), scientists produce preliminary reports after a research project is initiated. The work is then presented at seminars and colloquia; conference reports and proceedings are then produced and subsequently indexed in conference papers. A manuscript of the research paper is then sent to the publisher of a scientific journal, and preprints are distributed to researchers active in the area of research in question. Upon acceptance for publication by the publisher, the work appears in a list of manuscripts accepted for publication. The work is then published and then documented by scientific indexes and abstracts, then by annual reviews and then by citations in the literature. This model was found over the course of decades to be generally applicable across both the physical and social sciences (Crawford et al., 1996).

One of the key findings of the research conducted by Garvey and Griffith was that scientists tend to rely very heavily on informal communication channels in conducting research. In many fields of science, researchers in fact make comparatively little use of the scientific literature.

Scientists with common research goals, often separated by vast geo-graphic distances, frequently work together informally as a group. Price (1963) referred to such groups as "invisible colleges." A generation ago, this sharing of ideas and information typically involved informal meetings at conferences, visits to the workplaces of colleagues, tele-phone conversations and correspondence by mail. Invisible colleges play a crucial role in the communication of scientific information among researchers, particularly in the early stages of a research project prior to formal presentation at conferences or publication in a scientific journal. Today, informal communication within invisible colleges may include correspondence by e-mail and the use of digital libraries.

Garvey's model continues to be updated and refined. Hurd, for ex-ample, includes electronic journals, electronic conference reports, and contents databases in her modernized version of the communication cir-cuit. She emphasizes the importance of Big Science, which is not merely a change in scale, but is characterized by increased collaboration resulting in more multi-author papers and increasingly costly experi-ments (Crawford et al., 1996).

Models from Library and Information Science

A key focus for library and information science is the organization of literature and approaches to accessing it. Models developed for these purposes are models of scholarly communication; to look up a topic in an index to access a journal or monograph collection is a facet of schol-arly communication as are other informal acts examined by the model of Garvey. It is important to examine the organization and access of lit-erature as part of a larger scholarly communication circuit.

A common view of the organization of scientific literature is de-scribed in many standard bibliographic guides that classify the litera-ture according to primary, secondary and tertiary literature. Primary sources include letters of communication and articles in journals, the patent specification, various forms of conference literature, research re-ports and research papers. Secondary sources such as bibliographies, indexes, abstracts and current awareness services play the role of surrogation. Other secondary services such as dictionaries, directories, handbooks, etc., play the role of compaction of the information. The ter-tiary literature, including such sources as the bibliography of bibliogra-phies, directory of directories, and the guide to the literature, serve as a form of secondary surrogation of the information and to further dissem-inate the information (Gould, 1991; Subramanyam, 1981). While this model continues to serve as a basis for organizing and accessing the scientific literature, many of the approaches used in organizing and accessing literature within the framework of this model have undergone dramatic changes over the past century.

In the nineteenth century, the dominant controlled vocabularies were classification schemes. They were used to organize bibiliographic records in classified or classed catalogues. The terminology used in the catalogue was not user-friendly. The user could not find items listed by title. Title indexing was proposed and applied by Swanson Law in 1854 to produce an index to the titles in the *British Catalogue of Books* (Rowley, 1994). Charles Cutter objected to title-term indexing as titles may not express the true subjects of the works that they name, and works on the same subject would be separated if the phraseology of their titles were different. Through the work of Cutter, the alphabetical subject approach became widely adopted. The prototype of this approach was the *Library of Congress Subject Headings.*

Many of these basic approaches to organizing the literature were developed before the time of Garvey, prior to the information explosion and the introduction of electronic information retrieval devices. It was understood that if a scientist wanted information on a particular topic, the researcher would consult the tertiary literature to identify the appropriate indexes and other secondary literature that would in turn lead to all of the relevant primary literature on the topic. New ways of accessing the literature had to be devised to cope with the information explosion of the Second World War.

World War II produced an explosion in the growth of scientific and technical literature. The periodicals indexes at the time, which were either alphabetical or classified, were inadequate to accommodate the broad and complex emerging knowledge. It was clear that new methods of indexing and information retrieval, using machines, had to be found if scientists were to have any hope of keeping up with the growth of scientific and technical literature.

The application of Boolean principles to indexing and storing information began in the late 1940s and early 1950s. Mortimer Taube developed coordinate indexing, in which disciplines were broken down into single ideas with assigned index terms so machines could be used to organize and search information in the fields. In contrast with subject or classified indexes, which are hierachical, coordinate index terms are equal in weight (Smith, 1993). Many scientific and technical databases, such as MEDLINE and COMPENDEX, include subject headings in the form of descriptors from MeSH and SHE, respectively. In theory, relevant documents can be retrieved on the basis of the relevance judgements of indexers who have appropriate subject expertise.

Models of scholarly communication, as they are presented here, can leave the impression that scholarly communication is consistently becoming more efficient and effective as information technology is introduced into the scholarly communication circuit. Information is recorded in the literature and subsequently retrieved and passed on through communication channels at rapid speeds.

Problems begin to emerge, however, when we consider more precisely the intellectual content of the information that must pass from one scholar to another to produce effective scholarship. The intellectual content of documents relevant to scholarly discussions often far exceed that of the few superficial key words so often used to identify the documents we seek in the library. By examining the kinds of information contained in scientific documents that are necessary for scholars to do their work effectively, we can evaluate how successfully such essential information is communicated to scholars through literature. We can begin to consider what types of essential information contained in scientific documents do not make it through the scholarly communication circuit and explore the reasons why. This in turn provides a point of departure for considering how librarians can operate more effectively within the scholarly communication circuit.

SCIENTIFIC LITERATURE
AS FILTER, BARRIER AND BRIDGE
IN THE SCHOLARLY COMMUNICATION CIRCUIT

To understand what kinds of information must be transmitted through the scholarly communication circuit to enable effective scholarship to occur, it is necessary to examine how scientific documents are designed and used by scholars. Scientific documents have certain attributes designed to facilitate communication among scientists. It is to a large extent because of these attributes, or characteristics, that important information is either communicated to scholars, or lost. Scientific literature can therefore behave as a filter or barrier, or as a bridge in the scholarly communication circuit, depending on the circumstances. In this section, key features of scientific literature are presented and discussed, with an emphasis on how these features may result in important information being left behind in the scholarly communication circuit and never reach scholars for whom it was intended. Implications for librarians and information specialists are further discussed in a later section.

Social Process

The role of scientific documents as a part of social process in scientific scholarship is a factor that profoundly influences what information is of interest to scientists. If this fact is not understood by those who seek information for scientists, there is a significant potential for oversight. This brief discussion reveals one example of how the complex nature of scientific literature can significantly diminish the effectiveness of the scholarly communication circuit of scientists, despite a reliance on information technology.

Scholarly documents are more than artifacts of communication. A published scientific document embodies multiple scientific processes, such as the results of negotiations among teams of scholars as to the format to be used, the results that are to be conveyed and the interpretation of those results, the methodology, and who will receive credit for the work. The context of the document also conveys important information to the reader. An article published in the journal *Nature* is of much greater interest to a researcher than an article in the *Saturday Evening Post*; the reputation of the journal, the author and the funding agency may also have strong implications for how seriously a journal article is likely to be taken (Borgman, 2000).

Fleck (1979) identified various types of scientific literature based upon their social roles within scientific communities. Scientific journal articles, for example, are open for serious discussion among members of a small esoteric circle of scientists, but are of little interest to a much wider circle of amateur science enthusiasts and would never be published in a popular science magazine. Textbook science, on the other hand, is used for initiation into science.

According to Fleck, the epistemological state of a discipline (what is considered to be "fact" or what is considered to be "known" in a discipline) is given by the tertiary literature (e.g., handbooks, encyclopedias), not by journals. Due to the divergent points of view present in the primary journal literature it is neither appropriate nor possible to consider the tertiary literature as a neutral compilation of journal article content; a selection process is unavoidable. Journal articles merely present ideas for the consideration of other scientists in an evaluation process. The ideas can exist as facts only in the tertiary literature.

The secondary literature is also an expression of the social organization of science. The fact that certain journals and not others are selected to be included in indexes, for example, has a strong influence on how accessible a document will be to scientists. The fields that are indexed in secondary documents determine what types of information can be accessed and ultimately have the effect of selectively disseminating information.

A scholarly document represents competing ideas and theories indicative of certain schools of thought. Within a given field of science, a researcher is expected to adhere to certain norms. Senior scientists establish unwritten rules concerning what constitutes an issue or a problem that is worthy of being researched, what constitutes a valid hypothesis, or an acceptable methodology or paradigm in a particular field. A paradigm used in one field of science would often be completely unacceptable to many scientists in another. The results of a study of an ecosystem involving a complex mixture of parameters, for example, may be of interest to researchers in one field but in other fields may be viewed as having too many uncontrolled parameters to establish meaningful results. It is possible that an exhaustive search of the literature would not turn up any explicit mention of the issue; scientists simply adhere to the unwritten rules of their discipline. Often inference or deduction based on an intimate knowledge of scientific systems and processes carry far more weight in influencing a scientist's beliefs than the claims of a few new experiments by other researchers. An outsider could read exhaustively on a topic yet in the absence of informal communication with subject experts remain unaware of any disagreements between scientists on the matter, let alone the reasons why disagreements may occur. There is typically a range of dissent and disagreement that is tolerated within a scientific community. If a researcher violates certain norms, by drawing conclusions that break too many unwritten rules, for example, the researcher's work may not be accepted for publication. Scholarly documents are records of decisions made by scientists which establish accepted paradigms and norms in the conduct of research within scientific fields. As Agre (1995) points out, the literature is in a sense an ideology of both the documents and the institutions of research.

For a librarian, bibliographic attributes of documents such as the title and journal name may provide clues about the potential usefulness of a document to a particular researcher, but many of the crucial roles that documents play in the ongoing dialogs between scientists within a given subject domain are typically transparent to the librarian. These transparent roles are often the dominant factors that make particular documents either extremely valuable or useless to scientists.

Clearly, scientific literature can be an important filter in regulating the quality of scientific research through its influence on such social processes as peer review and dialog among scientists. It can also facilitate communication among scientists by providing efficient access to information. These social processes can also be viewed as barriers to communication. For example, there may be elements of censorship involved in the peer review process, and the selectivity of indexers may hinder the dissemination of certain information.

Representation of Meaning

A pervasive problem with using literature as a means of communicating information from one scholar or group of scholars to another has to do with the complex nature of language. Again, here is an example of aspects of scientific documents that have the potential to greatly diminish the effectiveness of the scholarly communication circuit, despite the enormous economic investments that governments and industry are making in information technology. Only with an awareness of the problem can we begin to address it.

A basic assumption behind the use of bibliographic tools such as indexes, abstracts and catalogues is that linguistic entities such as subject terms can represent the meaning of information contained in documents, as well as the information needs formulated in the user's queries. The question of how to represent concepts in a document accurately is problematic, however, because of the complexites of language.

A paradigm shift within linguistics, semiology (the theory of symbols) and epistemology occurred at the end of the nineteenth century and into the beginning of the 20th century (Chalmers, 1999). Earlier approaches were essentially *positivist* and the succeeding approaches were *structuralist*. The positivist view of language and meaning was that each constituent symbol of language had a one-to-one relationship of naming a unique "thing" or entity in the world. A word or symbol is a unique and objective identifier for a thing in the world. The positivist view maintains that the subjectivity and variability could be extracted from the symbolic representation.

Structuralism was based on the idea of a shared language that individuals would use as a basis for each utterance or action. It was difficult, however, to see how such a structure could be maintained uniformly, and the idea was not in keeping with observations of individuals' varying use of language. As a result, there was a move to poststructuralist theories of knowledge, one of which was hermeneutics.

The idea of meaning as a relative process, as a system of similarities and differences, was retained in hermeneutics but the subjectivity of interpretation was given far more attention than in structuralism. The meaning of a symbol or piece of information is considered to be unique to each individual. The meaning of a symbol or text depends on the interpreter's understanding in the moment. Understanding is itself created by interactions that occur through language, and these interactions in turn had to be understood through earlier understanding. A consequence of this circular process of seeing the part in the whole is that formal objective abstractions that describe some phenomena are brought down to the level of data by human interpretation.

The assumption that the context is not significant is true if one is attempting to retrieve all documents containing a given word, but false if one wishes to find the documents that are of greatest interest to a client. The assumption itself is neither good nor bad, but problems arise when there is a lack of awareness of it. "We should realise what assumptions our information representations are built on, and hence what they afford and what they inhibit" (Chalmers, 1999).

Cognitive approaches to information retrieval attempt to address the problem of subjectivity and the reduction process which Chalmers describes by creating a cognitive model of the user's world. This is done by a human intermediary who considers the problem of the user, or by an information retrieval system which represents the world of the user's information need, problem state, and domain work task based on information supplied by the user. Chalmers sees this approach as being a move towards positivism. Ingwersen (1996), however, argues that the cognitive approach to information retrieval is based on interaction between user and system or intermediary and the fact that the information flow between the two is associated with context, not just with the individual semantic value. Like Chalmers, Ingwersen emphasizes that any presuppositions, meaning and intentionality underlying the communicated messages are constantly lost, but that contextual clues are interpreted by the user to create a cognitive structure. The success of the printed KWOC (Key Word Out of Context) indexes was mainly due to their associative capabilities. A cognitive approach involves the exploration of many cognitive structures such as work task of the user, interest, social/organizational environment, as well as text/knowledge representations in order to provide clues for perception and interpretation.

Fragmentation of Scientific Literature

Another reason why the scholarly communication circuit of scientists may be substantially less effective than many people believe has to do with the fragmentation of scientific literature. An awareness of the problem is essential for information specialists if we are to begin addressing the most serious and profound problems inherent in scholarly communication.

One way in which scientists cope with the rapid growth of scientific knowledge is to specialize. A consequence of this ongoing trend is that connections between concepts and ideas are neglected and the scientific literature becomes fragmented. Kochen (1969, 1974) insisted on an urgent need to find ways of connecting these fragments of knowledge. According to Kochen, the real information explosion was an explosion of knowledge fragments and gaps between them. The literature fragments into specialty collections of nearly constant size as it grows and nearly every document contains some links to relevant prior documents which were used in its construction.

Swanson (1990, 1997) incorporated the ideas of Kochen and Popper into a method for discovering knowledge through database searching, and developed a trial-and-error database search process for creating new knowledge.

As Green (1995) points out, literature-based knowledge synthesis has been viewed as a phenomenon distinct from general literature retrieval, but "the distinction between knowledge synthesis and literature retrieval dissolves when the comparison is attempted first in the other direction, that is, by comparing literature retrieval to knowledge synthesis."

The idea of search strategy as a process or method of scientific inquiry was also explored in an article by Harter (1984). The strategy is based on producing many guesses and rejecting those that are incorrect. Online searching is viewed as an iterative process in which hypotheses are formulated, and iteratively tested, reformulated and retested.

Interaction with Information Retrieval (IR) Systems

The use of electronic information retrieval systems is almost universally relied upon in libraries. The strengths and weaknesses of such systems therefore merit very close consideration on the part of librarians and other information professionals. One could argue that this matter is potentially so serious that to ignore this issue is to ignore altogether the effectiveness of libraries and their role in the scholarly communication circuit of scientists.

Many users of World Wide Web search engines are concerned with finding a few relevant documents. Once these users have found the required information, they stop searching. In such cases, what is wanted is a system that finds a single relevant document and no non-relevant documents (i.e., a system that emphasizes precision). Research in the field of information retrieval more generally has tended to focus on users who want to find just a few relevant documents quickly (Wallis et al., 1996). In many situations, however, a user must be confident that a literature search has found all relevant material (i.e., high recall).

As Su (1994) showed, users in academic environments are more concerned with absolute recall (getting as many relevant documents on a topic as possible) than with precision (getting only some of the relevant information, but avoiding irrelevant information). This should not come as a great surprise; one would surely expect that scholars and academics would place high importance on being thorough and comprehensive in their research. Oversight of relevant scholarly documents could, after all, result in wasteful replication of experiments and a failure to keep abreast of important information.

From the perspective of scholarly communication, the comprehensiveness (or recall) of database searching is often so important that it can be viewed as a critical step in the scholarly communication circuit. Searches that fail to retrieve significant quantities of information potentially important to a researcher's work can be viewed as a disconnect in the scholarly communication circuit.

Some may suggest that there is no need for high recall retrieval systems and that existing tools are adequate. Cleverdon (1974) suggested that since there is significant redundancy in the content of documents, all the relevant information on a topic will be found in only one-quarter of the relevant material. Unfortunately one-quarter of the relevant material does not necessarily contain all the relevant information, so a user must find much more than one-quarter of the relevant documents to achieve high recall.

When one contemplates possible consequences of significant oversights, the idea of relying on a mysterious "black box" to retrieve information becomes especially troubling. How are we to know when we have retrieved all of the relevant information? How are we to judge whether we have retrieved all of it or only a miniscule fraction of the relevant information on a topic?

How Reliable Are Our Electronic Information Retrieval Devices in Uncovering Important Information?

As early as 1953 the lack of reliability of electronic information retrieval systems became apparent, when the first large scale experiment intended to test retrieval effectiveness was conducted. The test was in effect a contest between the novel, unconventional system devised by Taube, and the more conventional approach of the Armed Services Technical Information Agency (ASTIA) subject heading system. To evaluate the results, both contestants had to agree on which among the retrieved documents were or were not relevant to the questions. The test involved 98 questions applied to a test collection of 15,000 technical documents. Both teams agreed that 1,390 documents were relevant to at least one of the questions, but the contestants could not agree on the relevance of 1,577 other documents, an enormous disagreement that was never resolved (Swanson, 1997). A series of information retrieval experiments at the Aslib Cranfield Project in England in 1966 missed approximately 92% of the potentially relevant documents in the database used, according to Swanson.

At the UK Scientific Documentation Centre, with funding from the British Library, an extensive three-year study was conducted in the mid 1980s of bibliographic retrieval in science and technology to identify the most effective techniques of information retrieval. The study concluded that online searching was the least effective method of bibliographic retrieval. It was consistently outperformed by print sources by a wide margin (Davidson et al., 1988).

Why Are Information Retrieval Systems So Unreliable?

Information retrieval is at best problematic, largely because relevance judgements often tend to reflect subtle distinctions that cannot be discerned by anticipating word usage. The concept of relevance is inherently complex. It is frequently misunderstood and used ambiguously. The word "relevant" means different things depending on, amongst other things, which individual is asking, and the degree to which the individual is interested in finding all relevant material (Wallis et al., 1996). In an extensive review of the topic of relevance in the fields of documentation, information science and information retrieval, Mizzaro (1997) identifies many kinds of relevance and presents an approach to classifying relevance judgements along five dimensions: the kind of relevance judged; the kind of judge; what the judge can use for expressing his/her relevance judgement (surrogate, document, or information); what the judge can use for expressing his/her relevance judgement (query, request, information need, or problem); the time at which the judgement is expressed.

Many of the early evaluations of information retrieval systems (e.g., the ASTIA tests conducted in 1953, described earlier) involved third-party relevance judgements. It is unrealistic to expect a third party to find what the author of a query wants rather than what the author actually requests. In many cases users do not express their desires clearly and the same query can be given by two users with significantly different meanings.

Are Information Retrieval Systems Getting Any Better?

Many of the information retrieval systems used with large scientific bibliographic databases and that are commonly used in libraries have changed very little if at all in the past thirty years. The user must still rely on Boolean searching and concepts that are difficult to define are typically very difficult to search. Libraries typically are using the same technology that resulted in such poor test results in the 1960s. Much of the progress made by the IR research community has not been incorporated to any significant degree into IR systems in libraries. Information retrieval research has advanced over the years, and methodology has improved considerably as have results, but this can be seen much more inside IR research circles than in libraries.

Significant progress is evident in the TREC (Text Retrieval Conference) programme, which has established itself as the IR community's major evaluation exercise, involving many teams in many countries in a series of related tasks and tests (Sparck Jones, 2000). Here many of the best information retrieval systems in existence are tested and compared. Such systems involve relevancy ranking and through an on-going process of reformulating combinations of statistical keyword devices, have improved substantially over the past decade in their effectiveness.

Although the results of TREC are publicly available, they tend to be esoteric and much of the work tends to be easily misinterpreted. Results from the TREC project have been criticized, for example, for the way in which recall is calculated, raising questions about the true effectiveness of the IR systems being evaluated. These criticisms, along with outdated conclusions drawn over a decade ago by IR researchers, have led to pronouncements that IR effectiveness has reached an upper limit (e.g., Stoan, 1991; Sembok and van Rijsbergen, 1990; Salton, 1991; Norvig, 1994; Wilbur, 1996; Blair, 2002). As Saracevik et al. (2003) and Sparck Jones (2003) point out, however, many of these comments are now out of date and some of the comments are based on incorrect assumptions and misinterpretations of the literature. Saracevik and Sparck Jones discuss some of the key advances in information retrieval that have been made in the past ten years, and the evidence of it, in these papers.

This is all very well for the IR systems of TREC. Perhaps they are very effective, but how do most library IR systems compare with these? A common assumption held by many researchers in information retrieval is that natural language searching (i.e., the entry of search terms without Boolean operators and the relevance ranking of results) is superior to searching using Boolean operators. Hersh et al. (2001), however, recently showed that Boolean and natural language searching achieved comparable results, but the study involved professional database searchers. The fact that Boolean searching can be effective when performed by highly experienced searchers does not change the fact that it is an inherently complicated and difficult approach to information retrieval for the novice searcher, particularly when research involves poorly-defined problems.

It is surprising that the literature on information retrieval has for decades been filled with studies showing the intrinsically difficult nature of Boolean searching and the superiority of natural language searching, while vendors of major scientific databases continue to this day to offer Boolean searching as the only option for database searching. Furthermore, Boolean searching is being promoted to end-users. Since Boolean searching is so difficult to do well and involves such a long learning curve, should academic scholars be doing their own searching in the first place? Perhaps only highly experienced librarians should be performing database searches. What can IR research reveal about the ways in which IR systems should be used and who should use them? In the final analysis, only by considering the interaction between user and machine can we begin to bridge serious gaps in the scholarly communication circuit of scientists, which occur front and center in libraries and in machines that librarians encourage scholars to use.

User-Centered Aspects of Information Retrieval

The field of information retrieval developed along two rather distinct lines: the systems aspect of retrieval, and the user-centered aspect. The system-centered approach to IR has the longest history in IR research. It grew out of the problem of searching and retrieving relevant documents from IR. In the past two decades, however, a second research approach, that of users and intermediaries interacting with IR, has developed in IR. Although research in IR has been ongoing for over 30 years, most of the research and development in IR has concentrated on the improvement of effectiveness in automatic representation and searching, and has treated IR systems and processes as static and noninteractive. Research on the interactive aspects of IR has not reached maturity (Spink and Saracevic, 1997).

The problem of finding useful documents from an IR system consists of forming an understanding of a user's problem and translating that understanding into a query to be submitted to the information system. Saracevic et al. (1990) created a stratified interaction model of the human-human-computerized IR system based on comprehensive empirical studies of users interacting with computerized IR systems. The model considers interaction as a process involving two sets: user and system. A number of levels of interaction have been identified for both sets. On the surface level, dialogue is carried out by utterances and responses through interface with the system. At this level, the intermediary uses his/her knowledge of the system. On the cognitive level, users interact with texts and intermediaries clarify aspects of user modelling. On the situation level, the user interacts with a situation, and the need consequently leads to a result. During interaction, the deeper level changes can alter the surface level, and new search terms may be employed. There is therefore a direct interplay between deeper and surface levels. The interaction process is realized on the surface level while the effectiveness of the search terms and user judgements is established at the cognitive/situation levels (Spink and Saracevic, 1997).

In the studies of Saracevic et al. (1990), the authors state that prevailing theories do not conform with observed reality, due to a lack of real-life observations. Ellis et al. (2002) conducted a study of the information-searching behavior of academic researchers during mediated interaction with an IR system. The subjects were engaged in original research and were at different stages of the information search process. This meant that their required result and expectations were not similar. The study focussed on the interaction variables within Saracevic's (1990) model and involved data analysis of interview transcripts, on-line search results, and questionnaire results (Ellis et al., 2002). The interaction process helped the user obtain very useful results due to the intermediary's experience in defining search statements and reducing or refining outputs. This specific interaction aided the users significantly. The intermediary improved focus, completeness of retrieval, reducing nonrelevancy, and therefore increasing overall satisfaction. The study showed the critical importance of terminology in the search interaction. It dominated the search analysis results. The importance of feedback in the interaction process was highlighted, particularly with regard to the process of dealing with large sets of retrieved information. Traditional studies on feedback in IR concentrate on the role of the users' relevance judgements in search strategy formulation through automated relevance feedback techniques, but neglect helping users deal with the problem of magnitude (size of the sets retrieved). The authors noted that the problem or weakness with current IR systems is that they are made to answer well-defined questions, not uncertain ones.

IMPLICATIONS FOR INFORMATION PROFESSIONALS

The Importance of Social Process

As science and information technology continue to grow, the social dimension of scholarly communication will determine to a large extent how scientists seek and use information. It will also have a significant effect on the roles of the information professional. If molecular biologists as a scholarly community, for example, find that using a digital library to obtain copies of journal articles via fax does not meet their information needs, molecular biologists simply will not use that service if they can find a more practical alternative. If a scholarly community of scientists establishes through the peer review process that standards can be successfully met through informal communication channels and by browsing a few key journals, it is unlikely that many scientists within that community will spend their valuable time conducting extensive online searches of bibliographic databases. Certainly, all means of finding and obtaining information that are available to scientists in a given field will be viewed through a prism of social process. This fact provides a point of departure for information professionals in thinking about the provision of information services. Electronic resources may facilitate communication; they may provide fast and easy access to information. It is surely the information need of the scientist as determined by social processes operating within the scientific community, however, that will dictate how useful the resources and services of the library will be to the scientist, and ultimately how much they will be used.

The Importance of Librarian-Scientist Interaction

The information-seeking behaviors of scientists are based on perceptions of need and effectiveness. A librarian who understands the potential usefulness of good information may value it enough to spend hours looking through long lists of citations to uncover those few remaining precious gems. A scientist who is pressed to meet deadlines, however, may view a long and very comprehensive list of citations full of false drops as an indication that the librarian is incompetent, regardless of how many gems are present in the printout. A scientist might view a short printout with a few gems more favorably, but what happens to the reputation of the librarian if the scientist has asked for a comprehensive search and later uncovers all kinds of other gems via the scholarly communication circuit? A tiered printout with the most relevant postings in the first set, followed by other sets with what appear to be less relevant items might be a viable solution. Techniques such as those described by Harter (1984, 1990) can be effective, but sometimes results of Boolean searches are not so easily organized no matter what the searcher does, especially when stringent time constraints are involved. In any case, it's a sure bet that honesty and clear communication with the scientist will produce the best long-term results for the librarian, the library, the scientist and the scientific community alike. Providing the scientist with a short or medium-length printout of the results of a high-precision search along with an explanation of what must necessarily be involved in tracking down most of the other relevant material would help the scientist to understand the limits of information technology and give him/her a chance to make a reasonable decision about the true value of the service. This would surely be preferable to the scientist walking away with the impression that his/her valuable time is better spent looking for information elsewhere, or avoiding information altogether or resorting to end-user searching.

Such user-intermediary interactions do not take place when a scientist decides to use electronic resources remotely and finds the results lacking. E-mail, online chat or videoconferencing could be used to provide reference service.

Scientists can uncover valuable information by doing their own database searches, but often do not have the skills to cope with large posting sets full of false drops. The effectiveness of end-user searching is questionable in view of the fact that good online search results are often difficult for experienced information professionals to achieve, and in view of what the studies of Ellis et al. (2002), described above, have to say about the importance of the intermediary in the search process. Here again, the value of face-to-face communication between librarian and scientist is clear. Encouraging scientists to do their own searching while urging them to have a librarian do follow-up searching may be one solution to the limitations of end-user searching. On the other hand, scientists often have a clear-cut, well-defined information need, in which case the query necessary to achieve high recall is often simple to formulate.

When the problem being researched is less well-defined, the end-user often has little sense of when or how to formulate complex queries necessary to achieve high recall, or of how to deal with large posting sets.

Verification of Scholarly Communication Models: The Information Audit

Models of information flows can be very useful for planning. Textbook models of scholarly communication circuits (e.g., Gould, 1991; Crawford et al., 1996) provide a good point of departure in the designing of information services. There is often a range of information-seeking behavior that scientists working within a field are not typically shown in such models. Palmer (1991) showed that personality can be strongly related to the information-seeking behaviors of scientists. Using psychometric testing and cluster analysis she was able to place them into categories ranging personal information-seeking style on a continuum from "active" to "passive." She identified one group of scientists in particular, her "type 5G3" which ". . . needed looking after," and pointed out that it would be worthwhile for a librarian to be able to distinguish this type in a research environment.

Other types of classification are of dubious value. Abel (1991) suggests that scientists who are in the best positions to effectively promote the library are often leaders in their field. They obtain all of the information they need through the invisible college and by-pass the library completely. He maintains that the vast majority of researchers in a field do rely on library resources and services, but they do not have the prestige of these leaders of the invisible college. As Abel points out, these leaders are often the very individuals that presidents and P.R. departments of institutions like to target for marketing purposes when they want to raise the visibility of the institution. He describes a rule of thumb among consultants to both non- and for-profit organizations: 80% of an organization's costs are incurred by 20% of the organization's products or activities. The idea under consideration here is to expend an inordinately large proportion of resources on small but powerful elite groups to gain their support, even at the expense of the majority. He then poses the question, "Is the extremely high cost of providing additional service at the margin worth the reward of winning the movers and shakers back to an increased use of their local library?" The author then suggests that the "movers and shakers" are already well taken care of by the invisible college and that the gains from collaborating with "synthesizing scholars, teachers, and students" who are not in his view "movers and shakers" might well outweigh the losses.

The author may be exaggerating the difficulties involved in addressing the needs of first-rate scientists. Consider the example of a molecular biologist who had published numerous times in the journal *Nature* and had published over 150 peer-reviewed journal articles. True to the established model of molecular biologists' scholarly communication circuit (Covi, 1999), he obtained his information from the following sources: five key journals for which he owned personal subscriptions, conferences, and informal communication with colleagues. In addition to these sources, a librarian performed an online search for him once a year (that was not in the standard model of Covi). One clear information need emerged from an interview with the scientist: reliable patent searching (that was not in the standard model either). Patent searches had been performed for him at each of the universities where he had worked. Nevertheless, the U.S. Trademark and Patent Office rejected many of his applications when they retrieved patents from their databases showing the existence of prior art. Here was a critical information need that was not being met. Would making arrangements for a few reliable patent searches really constitute an ". . . extremely high cost of providing additional service at the margin"?

The less scientists and librarians interact, the less opportunity there is for librarians to become aware of the real needs of scholars. There is a danger of overlooking the most critical, pertinent needs of scientists by relying too much on stereotypes and on general models of information-seeking behaviors. With the development of, and reliance upon, digital libraries comes the risk of diminished interaction between scientist and librarian. As Lancaster et al. (1997) warned, there is a strong possibility that libraries will become mere electronic switching stations for scholars.

Of course one-on-one interaction with scientists may be an inefficient way of learning about the needs of scholars. On the other hand, a few good information audits of the needs of targeted scholars can send strong messages through the informal scholarly communication circuit. When this happens, scholars refer other researchers to the librarian. Surveys may help improve efficiency in gaining information, but they will inevitably miss much that will emerge from one-on-one interactions that can be helpful in the design of information services and resources such as digital libraries.

Digital Libraries

Interest in digital libraries has expanded significantly in the past few years. As Borgman (1999) notes, the availability of research funding under this term has attracted scholars and practitioners from a wide variety of backgrounds, some of whom have minimal prior knowledge of related areas such as information retrieval, and sometimes research projects are simply relabeled "digital libraries." What exactly are digital libraries? In view of the issues regarding information retrieval discussed earlier, what can and should be done to offer scholars assistance in using digital libraries?

There are sharply contrasting definitions of digital libraries in various literatures. Borgman (1999) notes two complementary research-oriented definitions of the term "digital libraries" developed at the NSF-sponsored Social Aspects of Digital Libraries workshop (Borgman, 1999):

1. Digital libraries are a set of electronic resources and associated technical capabilities for creating, searching and using information. In this sense they are an extension and enhancement of information storage and retrieval systems that manipulate digital data in any medium (text, images, sounds; static or dynamic images) and exist in distributed networks. The content of digital libraries includes data, metadata that describe various aspects of the data (e.g., representation, creator, owner, reproduction rights) and metadata that consist of links or relationships to other data or metadata, whether internal or external to the digital library.

2. Digital libraries are constructed, collected and organized, by (and for) a community of users, and their functional capabilities support the information needs and uses of that community. They are a component of communities in which individuals and groups interact with each other, using data, information and knowledge resources and systems. In this sense they are an extension, enhancement and integration of a variety of information institutions as physical places where resources are selected, collected, organized, preserved and accessed in support of a user community. These information institutions include, among others, libraries, museums, archives and schools, but digital libraries also extend and serve other community settings, including classrooms, offices, laboratories, homes and public spaces.

Borgman (1999) noted that researchers working in the fields of computer science and/or engineering focus on digital libraries as content collected on behalf of user communities, while librarians focus on digital libraries as institutions or services. Marchionini and Fox (1999) state that "digital library work occurs in the context of a complex design space shaped by four dimensions: community, technology, services and content." They emphasize that there is a great need for attention to reference and question-answering, on-demand help, fostering of citizenship and literacy, and ways of simplifying involvement of user communities. In the March 1994 Digital Library Workshop, a definition of a digital library emphasized that a full service digital library must accomplish all the essential services of traditional libraries (Gladney et al., 1994). Many other definitions of digital libraries mention the services aspect either explicitly or implicitly (Oppenheim and Smithson, 1999; Rusbridge, 1998 (http://www.dlib.org/dlib/july98/rusbridge/07rusbridge.html); Waters (http://www.clir.org/pubs/issues/issues04.html#dlf), 1998). According to Sloan (1998), "technology and information resources, on their own, cannot make up an effective digital library."

Covi (1999) suggested that subject specialists help users formulate disciplinary search strategies and provide assistance in developing new resources. Arms (2000) suggests that the skills of librarians may be required in the case of complex information searches, but believes that there have been significant improvements in the user-friendliness of electronic databases provided by electronic database search services and Web search engines for end-users. In Arms' judgement, the need for human intermediaries in information searching has been reduced significantly.

A recent study performed by Cooper (2001) analysed the usage of the Melvyl Web catalogue at the University of California at Berkeley. About 7.4 million search statements were executed over the study period covering 479 days. Users spent about 3 minutes per search in the Medlars database and about 2.2 minutes per search for the catalogue database. One of the major problems observed was the absence of human intermediaries and inadequate online support for query formulation and modification.

As a component of the scholarly communication circuit, digital libraries can be viewed as a collection of electronic resources such as electronic databases and websites. It can in fact be viewed as just another term for the electronic resources that are already in place at libraries. In fact, Borgman (1999) poses the question of whether a database is a digital library. Her conclusion is that:

Some portion of electronic databases on the Internet, on proprietary systems and on CD-ROMs are digital libraries in the senses defined by the research community. On a case-by-case basis we can judge the degree to which given databases are organized collections, whether they were created for a specific community and whether their capabilities are sufficient to distinguish them from other forms of information retrieval systems, for example.

Whether a resource is called a database or website or digital library does not change the fundamental issues that have been discussed, such as what constitutes relevance, and how a scholar can identify as much of the relevant scholarly information on a particular topic as possible. One of the key issues for librarians is how reference services can be incorporated into the digital library to address these crucial issues effectively.

Bridging Gaps and Filtering Information in the Scholarly Communication Circuit

As shown earlier, gaps in the flow of information through the scholarly communication circuit occur in many ways. Where one solution to bridging a gap may be found, other factors may interact and create other gaps. For example, Swanson and Smalheiser (1997) describe an interactive system for finding complementary literatures which can aid in scientific discovery. If a librarian could use such a system to uncover new important avenues of research, it may change nothing if scientists are not receptive to the idea. Since individuals within a field vary in their information-seeking behaviors, it may be necessary for a librarian to identify the specific individuals who might be receptive to the new approach. It may therefore be necessary to work on several different levels simultaneously to significantly improve the flow of information in the circuit.

The problems associated with finding documents relevant to a scientist's information need based on matching topics is a basic dilemma in information science. Based on the discussions here, it would appear that this can be best accomplished by an intermediary working with a scientist; only the scientist is in a position to accurately judge relevance of a document to his/her needs, yet Boolean searching is not a strong skill of many scientists. There are of course other approaches to information retrieval which rely on other kinds of relationships, such as citation analysis. Chalmers (1999) suggests the possibility of using collaborative filtering. An example of this would be Amazon.com's book recommender which treats each book purchased as an interpretive act that adds to a purchaser's profile. Profiles of different individuals are compared. Chalmers also suggests a number of other approaches to accessing information, including a path model in which the use of symbols such as words, URLs, and filenames displayed in a web browser are logged over time. At any point in time, the word or token nearest to the cursor is logged, and the log information is processed to reduce redundant logging. Patterns of symbol recurrence determine relevance and information need. The most recent sequence of path entries is treated as an implicit request for recommendations. The system tallies the symbols collected from these windows, and with some processing of the symbols presents the symbols as a ranked recommendation list. Each path is visible to anyone who contributes paths, so the set of paths is treated as a shared resource.

Hopkins (1995) notes that Garfield, in the early 1970s, envisioned a time when citation databases in electronic form would assist librarians in identifying "key" papers for their patrons. The idea was to develop databases in which one could input a subject descriptor or descriptors into an electronic form and have the system generate a ranked list of the most cited sources in a particular area or field of study. As the author notes, apparently Garfield can do this internally at ISI, but this service apparently has not yet been made more widely available to subscribers of the databases.

Hopkins also reports an approach to quality filtering in which the abstracting of the journal's contents in *Abridged Index Medicus*, a subset of journals selected from the full *Index Medicus* on the basis of quality, is used in combination with the *Science Citation Index* Impact Factor.

On a small scale, one can proceed to compare results of two or more methods of searching for information, and to examine how they can be used in tandem. One can examine on a small scale the information needs of clients by conducting in-depth information audits. One can conduct extensive reviews of the literature on various methods used to access and retrieve information. Only by considering the day-to-day work of helping scientists in the context of the overall scholarly communication circuit, however, can one be confident of the usefulness of one's efforts.

CONCLUSION

The scientific literature plays important roles in social processes within scientific communities. Judgements that scientists make regarding the validity, reliability, value, and importance of a scientific document are often based on criteria that are not evident to the information professional, such as how consistent the information is with his/her knowledge of the subject and understanding of the ongoing dialog among scientists. A published scientific document is the embodiment of social processes, and shows the results of negotiations among teams of scholars regarding such matters as the format to be used, the results that are to be conveyed and the interpretation of those results, the methodology, and who will receive credit for the work. Attributes such as author, methodology and results often have different meanings for the scientist than they have for the information professional. An information professional can often identify the leading subject authorities in a field through recent review articles and citation indexes, but some scientists may place as much value on the work of lesser-know researchers based on certain criteria such as thoroughness or the ways in which authors frame issues and write hypotheses. This observation supports the view of researchers in information retrieval that, ultimately, only the user (or user working with an intermediary) can make valid decisions concerning the relevance of a document to the user's information needs.

"Relevance" of literature is conceptualized in information retrieval as a correspondence between a text segment and a user need that it may help. Topicality is a major factor in establishing that correspondence, but there is little real understanding of how topics of a user need and relevant topics of text segments relate. Traditional retrieval theory and traditional retrieval systems assume that topic matching can be used to identify relevant documents, but topic matching for this purpose has been found to be wanting. This is due in part to the fact that relevance judgements often tend to reflect subtle distinctions that cannot be discerned by anticipating word usage. The limits to information-retrieval effectiveness based on topicality may also be due to the assumption that matching relationships are the only relationships that count. It is also due in part to the loss of information that occurs when meaning is represented in indexes and abstracts.

Most communication among scientists occurs informally through e-mail, discussion at conferences, telephone conversations, and other informal channels. Communication and information-seeking behaviors are strongly influenced by social processes within scientific communities. They depend on the scientific discipline in question and to some extent on the personality of the individual.

An important issue for many scientists engaged in scholarly research is the comprehensiveness of the literature search. Overlooking important scientific documents can result in wasteful duplication of the research of others, and may in effect prevent a scientist from accessing information that is crucial to doing meaningful research. Comprehensiveness of the literature search is more important in some fields of science than in others. For a computer scientist it may be far more important to know about a few current ideas that could have many different applications than to spend a great deal of time conducting a comprehensive review of the literature. It is crucially important to retrieve all the relevant or partially relevant documents when lodging a new patent at the patent office; duplication of effort as a result of overlooking a patent could result in losses in the millions of dollars for a company.

Comprehensiveness is an important aspect of scholarly research in many fields of science, so academic researchers are generally much more concerned about high recall than the general World Wide Web user population. It is a potential barrier to the reliance on many new approaches to information access, such as using collaborative filtering. It is usually difficult to achieve high recall using Boolean searching, especially when the information need of the scholar is not clearly defined. In such situations the role of the intermediary in translating user needs into queries is very important. End-user searching can be effective if the researcher only wants to find a few good documents on a topic, but the need to conduct comprehensive searches is frequently problematic for the end-user. The end-user frequently has little awareness of how much relevant information is likely to be overlooked in database searches. In essence, this oversight can amount to a significant gap in the scholarly communication circuit of scientists. The information professional can play a vital role in bridging that gap. Research in the field of information retrieval has significantly improved the effectiveness of information retrieval systems, but for the most part these improvements have yet to be incorporated into the retrieval systems used for large scientific databases. Significant advances in the field of information retrieval have essentially gone unnoticed by information professionals in general, partly because these advances are not evident in the systems that are still used in libraries, and partly because much of the literature on information retrieval trends has been widely misinterpreted or misunderstood. Librarians may be able to play a role in bridging the gap between information retrieval research and library systems development. Advances in information retrieval technology will not be incorporated into new library systems unless market competition allows that to occur. Librarians can work in cooperation with researchers in information retrieval, with a view to evaluating systems on the basis of retrieval effectiveness. Greater selectivity in the purchase of new systems could help create market conditions that would encourage the development and availability of systems that use modern IR technology.

The importance of introducing information retrieval systems that rely on advanced and modern information retrieval technology has been raised to new levels by the high visibility of digital libraries. With the introduction of the World Wide Web, scientists once served by professional online searchers were encouraged to find their own information as end-users. With the introduction of digital libraries they are now at risk of being left to fend for themselves using IR systems that are difficult at best for information professionals to use, let alone end-users. Steps can be taken to address the difficulties and limitations of the systems that are being used in digital libraries. These may include offering online chat sessions to assist users of IR systems, outreach programs including targeted visits to offices of selected scholars, and well-advertised services offering online searches conducted by librarians.

The fragmentation of the scientific literature due to specialization leaves a wealth of undocumented and undiscovered relationships between ideas. Approaches to literature searching have been developed to identify such relationships and hence create new knowledge. The fragmentation of scientific knowledge is a highly neglected aspect of science. There is a wealth of opportunity for librarians to become actively engaged in the process of creating new knowledge through such approaches. There may, however, be a significant social barrier to such approaches due to a lack of motivation or interest on the part of scientists. It would be wise to make scientists aware of such approaches through demonstrations.

The scientific literature can be accessed effectively with information technology. A lack of awareness of the complexities of the scientific literature, however, combined with blind faith in the effectiveness of information technology, often result in significant gaps in the scholarly communication circuit of scientists.

REFERENCES

Abel, Richard. 1991. Invisible colleges, information, and libraries. *Library Acquisitions: Practice & Theory* 15: 271-277.

Agre, Philip E. 1995. Institutional circuitry: thinking about the forms and uses of information. *Information Technology and Libraries* 14: 225-230.

Blair, David C. 2002. Some thoughts on the reported results of TREC. *Information Processing and Management* 38: 445-451.

Borgman, C.L. 1999. What are digital libraries? Competing visions. *Information Processing and Management* 35: 227-243.

_____. 2000. Digital libraries and the continuum of scholarly communication. *Journal of Documentation* 56, no. 4: 412-430.

Chalmers, Matthew. 1999. Comparing information access approaches. *Journal of the American Society for Information Science* 50, no. 12: 1108-1118.

Cleverdon, C.W. 1974. User evaluation of information retrieval systems. *Journal of Documentation* 30, no. 2: 170.

Cooper, M.D. 2001. Usage patterns of a Web-based library catalog. *Journal of the American Society for Information Science and Technology* 52, no. 2: 137-48.

Covi, Lisa. 1999. Material mastery: situating digital library use in university research practices. *Information Processing and Management* 35: 293-316.

Crawford, Susan Y., Julie M. Hurd, and Ann C. Weller. 1996. *From print to electronic: the transformation of scientific communication*. Medford, N.J.: Information Today, Inc.

Davidson, P.S., McDonagh, J.D.P. Meldrum, and A. Moss. 1988. *International bibliographic review on costs and modelling in information retrieval: Report on completion of data collection, assessment of the efficiency of searching, and on means of indexing this information*. London: British Library.

Ellis, David, T.D. Wilson, Nigel Ford, Allen Foster, H.M. Lam, R. Burton, and Amanda Spink. 2002. Information seeking and mediated searching. Part 5. User-intermediary interaction. *Journal of the American Society for Information Science and Technology* 53, no. 11: 883-893.

Fleck, L. 1979. *Genesis and development of a scientific fact*. Chicago, IL: The University of Chicago Press.

Garvey, W.D. 1979. *Communication: The Essence of Science*. Elmsford, NY: Pergamon Press.

Garvey, W.D., and B.C. Griffith. 1972. Communication and information processing within scientific disciplines: Empirical findings for psychology. *Information Storage and Retrieval* 8: 123-126.

Gladney, H.M., E.A. Fox, Z. Ahmed, R. Ashany, N.J. Belkin, and M. Zemankova, 1994. *Digital library: gross structure and requirements, report from a March 1994 workshop*, available at: www.csdl.tamu.edu/csdl/DL94/paper/fox.html (accessed January 8, 2003).

Gould, Constance G., and Karla Pierce. 1991. *Information needs in the sciences: An assessment*. Mountain View, CA: The Research Libraries Group.

Green, Rebecca. 1995. Topical relevance relationships. I. Why topic matching fails. *Journal of the American Society for Information Science* 46, no. 9: 646-653.

Harter, Stephen P. 1984. Scientific inquiry: a model for online searching. *Journal of the American Society for Information Science* 35, no. 2: 110-117.

_____. 1990. Search term combinations and retrieval overlap: a proposed methodology and case study. *Journal of the American Society for Information Science* 41, no. 2: 132-146.

Hersh, W., A. Turpin, S. Price, D. Kraemer, D. Olson, B. Chan, and L. Sacherek. 2001. Challenging conventional assumptions of automated information retrieval with real users: Boolean searching and batch retrieval evaluations. *Information Processing and Management* 37, no. 3 (May): 383-402.

Hopkins, Richard L. 1995. Countering information overload: the role of the librarian. *The Reference Librarian* 49/50: 305-33.

Ingwersen, Peter. 1996. Cognitive perspectives of information retrieval interaction: elements of a cognitive IR theory. *Journal of Documentation* 52, no.1: 3-50.

Kochen, M. 1969. Stability in the growth of knowledge. *American Documentation* 20 no. 3: 186-197.

Kochen, M. 1974. *Integrative mechanisms in literature growth.* Westport, CT: Greenwood Press.

Lancaster, F.W., and Beth Sandore. 1997. *Technology and Management in Library and Information Services.* Champaign, Ill.: University of Illinois Graduate School of Library and Information Science.

Mizzaro, Stefano. 1997. Relevance: the whole history. *Journal of the American Society for Information Science* 48(9): 810-832.

Norvig, P. 1994. Review of text-based intelligent systems. *Artificial Intelligence*, 65: 181-188.

Oppenheim, C. and D. Smithson, 1999. What is a hybrid library? *Journal of Information Science* 25, no. 2: 97-112.

Palmer, Judith. 1991. Scientists and information: I. Using cluster analysis to identify information style. *Journal of Documentation* 47, no. 2: 105-129.

Popper, K.R. 1959. *The logic of scientific discovery.* New York: Basic Books.

_____. 1979. *Objective knowledge: an evolutionary approach.* Oxford: Oxford University Press.

Price, D.J. de Solla. 1963. *Little Science, big science.* New York: Columbia University Press.

Rowley, Jennifer. 1994. The controlled versus natural indexing languages debate revisited: a perspective on information retrieval practice and research. *Journal of Information Science* 20, no. 2: 108-119.

Rusbridge, C. 1998. Towards the hybrid library. *D-Lib Magazine*, available at: http://www.dlib.org/dlib/july98/rusbridge/07rusbridge.html (accessed 2 January 2003).

Salton, G. 1991. Developments in automatic text retrieval. *Science* 253, 974-980.

Saracevic, T., H. Mokros, and L. Su. 1990. Nature of the interaction between users and intermediaries in on-line searching: A qualitative analysis. *Proceedings of the Annual Meeting of the American Society for Information Science* 27, 47-54.

Saracevic, Tefko, E. Voorhees, and D. Harman. 2003. Letters to the editor. *Information Processing and Management* 39: 153-159.

Sembok, T.M.T., and C.J. van Rijsbergen. (1990). Silol: A simple logical-linguistic document retrieval system. *Information Processing and Management* 26, no. 1: 111-134.

Sloan, B.G. 1998. Service perspectives for the digital library remote reference services. *Library Trends* 47, no. 1: 117-43.

Smith, Elizabeth S. 1993. On the shoulders of giants: from Boole to Taube: the origins and development of computerized information from the mid-19th century to the present. *Information Technology and Libraries* 12, 217-226.

Sparck Jones, Karen. 2000. Further reflections on TREC. *Information Processing and Management* 36: 37-85.

Sparck Jones, Karen. 2003. Letters to the editor. *Information Processing and Management* 39: 153-159.

Spink, A. and T. Saracevic. 1997. Interaction in IR: Selection and effectiveness of search terms. *Journal of the American Society for Information Science* 48, no. 8: 741-761.

Stoan, Stephen K. 1991. Research and information retrieval among academic researchers: implications and library instruction. *Library Trends* 39, no. 3: 238-57.

Su, L.T. 1994. The relevance of recall and precision in user evaluation. *Journal of the American Society for Information Science* 45, no. 3: 207-217.

Subramanyam, Krishna. 1981. *Scientific and technical information resources.* New York: Marcel Dekker, Inc.

Swanson, Don R. 1990. Integrative mechanisms in the growth of knowledge: A legacy of Manfred Kochen. *Information Processing & Management* 26, no. 1: 9-16.

_____. 1997. Historical Note: Information retrieval and the future of an illusion. In *Readings in Information Retrieval*, San Francisco, CA: Morgan Kaufmann Publishers.

Swanson, Don R. and Neil R. Smalheiser. 1997. An interactive system for finding complementary literatures: a stimulus to scientific discovery. *Artificial Intelligence* 91: 183-203.

Wallis, Peter, and James A. Thom. 1996. Relevance judgements for assessing recall. *Information Processing and Management* 32, no. 3: 273-286.

Waters, D.J. 1998. What are digital libraries? *CLIR Issues*, 4, available at: http://www.clir.org/pubs/issues/issues04.html#dlf (accessed 2 January 2003).

Scholarly Communication in Flux: Entrenchment and Opportunity

Kate Thomes

SUMMARY. Science and technology librarians know from direct experience about escalating journal prices, restrictive license agreements, and the commercialization of scholarly publishing. While digital technologies offer exciting new opportunities for rapid, broad dissemination of scholarship, the scale of the scholarly communication system as a whole makes change slow and difficult. This paper cites the varied stakes of faculty, publishers, and librarians as partial explanation for this entrenchment. Despite some resistance, significant national and international initiatives do present a variety of new possibilities for the culture and technology of scholarly communication. This paper suggests several activities for librarians to help effect change at the local level. It also recommends creation of an Access Impact rating scale to assess the effect of copyright transfer agreements on the availability of scholarship. *[Article copies available for a fee from The Haworth Document Delivery Service: 1-800-HAWORTH. E-mail address: <docdelivery@haworthpress.com> Website: <http://www.HaworthPress.com> © 2002 by The Haworth Press, Inc. All rights reserved.]*

KEYWORDS. Scholarly communication, open access, librarians' roles, copyright retention, copyright transfer, access impact

Kate Thomes, BA, MA, is Head, Bevier Engineering Library, University Library System, University of Pittsburgh (E-mail: kthomes@pitt.edu).

[Haworth co-indexing entry note]: "Scholarly Communication in Flux: Entrenchment and Opportunity." Thomes, Kate. Co-published simultaneously in *Science & Technology Libraries* (The Haworth Information Press, an imprint of The Haworth Press, Inc.) Vol. 22, No. 3/4, 2002, pp. 101-111; and: *Scholarly Communication in Science and Engineering Research in Higher Education* (ed: Wei Wei) The Haworth Information Press, an imprint of The Haworth Press, Inc., 2002, pp. 101-111. Single or multiple copies of this article are available for a fee from The Haworth Document Delivery Service [1-800-HAWORTH, 9:00 a.m. - 5:00 p.m. (EST). E-mail address: docdelivery@haworthpress.com].

10.1300/J122v22n03_09

INTRODUCTION

For many science and technology librarians the system of scholarly communication is something we work with daily yet may feel little influence over. We understand the beauty of library cataloging and classification, of abstracting and indexing services, of consistently published, numbered and titled journals, and of the peer review system. It can be a truly marvelous feeling to help a scholar retrieve an article that is old, obscure, or partially cited with the full understanding of why it is retrievable, and what it has taken to construct a system that tracks generations of scholarship. Yet most practicing librarians today inherited this mature system and our role has been to work within it and master its nuances. Many of us have lost sight of our role as creators of a system that ensures access to and retrieval of the scholarly record.

The system was designed to foster the dissemination of information, to register claim to ideas and insights, to provide a means of evaluating scholarship, to edit, format and distribute articles, and to keep and maintain that scholarship for future use.[1] Various sub-systems developed over time and by the end of the 20th century faculty, publishers, and librarians clearly understood their roles. For a time the roles and goals of each of these groups were basically in balance. However, evolution, adaptation, and change inevitably worked on all three, over time revealing differences of purpose and motivation. The chief change-agent has of course been digital technology. This technology offers wonderful new opportunities for faculty, publishers, and librarians but crystallizes differences in the stakes held by each group.

The exquisite paper by Guédon,[2] reports from the Pew Higher Education Roundtable[3] and the Association of Research Libraries[4] and the Mellon Foundation[5] provide detailed perspective on the system's history and can be considered essential reading on scholarly communication in the late 20th century. During the past few years, discussion on scholarly communication has moved from raising awareness of the "serials crisis" and its negative impact on academe, to the clear and pervasive understanding that the current system is untenable, and to the search for creative new means for accomplishing academic goals that remain valid. Presenters at meetings and workshops of national library organizations such as the American Library Association (ALA) and the Association of Research Libraries (ARL) no longer detail this history, assuming that attendees understand the issues. The question now is not whether we need change but how change will occur and what form it will take. Rather than restating the origins of the problem, which are so richly documented,[6] this article reviews some of the cultural challenges we face in dealing with a complex and entrenched system, reviews the perspectives of key stakeholders, and suggests some specific activities librarians can undertake to assist the change process. In this discussion I will draw on my experience with faculty and publishers as an engineering librarian over the past 10 years, who like all my colleagues has needed to analyze and respond to the evolving scholarly communication system.

In biological systems change does not occur overnight with the appearance of a full-blown new organism, but occurs incrementally, with small advances and retreats in a variety of areas. I believe this is where we are today in academe: a fertile and creative environment. Scholarly publishing is currently a "hot-topic" with interesting articles and new initiatives appearing daily. Listservs, conferences, and workshops continually discuss new developments and perspectives related to scholarly communication. For example, the Free Online Scholarship weblog[7] is an excellent source for current awareness on a broad range of topics related to scholarly communication. The quantity of all this information is challenging to keep up with.

As individual academic librarians in reference or middle management, it can be difficult to see where we have input to this process; in much the same way individual faculty may not see their roles in the big picture. However, by understanding the various stakes and pressures on the system as a whole and by maintaining awareness of new initiatives, I believe we position or prepare ourselves to make important professional contributions by engaging faculty and other colleagues in informed discussion of these issues.

WHY IS THE SYSTEM SO HARD TO CHANGE?

Simply put, it is large, complex, and abstract. It is based on longstanding cultural and sociological practices within academe. Librarians, faculty, and publishers have all made significant contributions to the print system of scholarly communication, each attending to their own responsibilities and interests. Faculty, publishers, and librarians see the system very differently from each other and many do not see it as broken. Inertia plays a significant role in maintaining the status quo for those who feel the system works effectively or are fearful of disrupting the current system before a new one is clearly and fully in place. Stakeholders who do see opportunities for change in the digital arena are working to create a new system that fosters their own goals. If this is done in isolation, without thinking of the system as a whole, we will simply carry the problems of the past with us into the future.

As science and technology librarians, we experience the pressures and tensions between these interests regularly in our work. A brief and selective look at some of these viewpoints, drawn from the literature as well as from personal experience, may help illustrate challenges we face on a daily basis.

Faculty Views

As the creators of scholarship, faculty are the central players in this system. Publishers and librarians work to facilitate different stages of the communication process, but do so in response to the productivity and needs of faculty.

Faculty focus on issues related to the content of the science and technology itself, they understand the intellectual and hierarchical structures within their disciplines, and are keenly attuned to status variations between sub-fields of study, institutions, research groups, and publications. For faculty what counts is unfettered access to recent (as well as historical), high quality scholarship with which they can work to advance their own research. Faculty want a recognizable mechanism for peer review, easy identification of relevant literature, and quick, convenient, and free access to research articles.

Publications are the currency of their professions. Yet they reap no direct financial benefit from journal sales. In fact, faculty view publishing as part of the gift culture in academe and eagerly give their scholarship to the world to advance discourse.[8] Where and how often they publish directly affects their career security and opportunities, however junior faculty often do not feel they can afford to do anything but work with the system as it is until their reputations and careers are established.

Many faculty see the current system of scholarly communication as an effective, known, and reliable system that is not broken and therefore does not need to be fixed. Many other faculty see difficulties with the current system and have been early adopters of alternative publishing venues.

Publisher Views

Publishers come in different stripes and do not all share a single "publisher's mission statement." While faculty and librarians can claim to serve scholarship in a pure sense, publishers, arguably, serve two masters: academe and the market economy. Professional and scholarly society publishers align themselves more closely with academe, although they need to financially break even or make a modest profit in order to continue their service. Commercial publishers, on the other hand, are clearly businesses operating in the global economy and using scholarship as the raw material of their "product."

Commercial and society publishers have played a significant role in creating a stable, functional system of scholarly communication in the print environment and are developing many useful and creative innovations in the emerging digital environment. For commercial publishers these innovations are researched and designed in a business environment, rather than an academic environment, utilizing business models for market research and product development. Their bottom-line stake is market share and revenue stream.

Librarian Views

Librarians collect what has been produced, organize it, and make it accessible to the scholarly community. We collect broadly, across all disciplines. We maintain awareness of research trends and faculty interests, and are also keenly attuned to the publishing practices and pricing changes of both commercial and society publishers. Librarians see on a daily basis the need for systematic information organization and retrieval mechanisms that have long-term viability. As a professional value, librarians endorse practices that promote open, barrier-free access to scholarship, and share increasing concern over issues of copyright transfer.[9]

Common Ground

It is the tension exemplified by goals of faculty and librarians versus the goals of commercial publishers that underlies much of the pressure to change the system of scholarly communication. Is scholarship a social good that ought to be made available to scholars of the world at no or minimal costs? Is scholarship a commodity appropriately offered for sale in the market economy? These are political, ideological, and economic issues that need to be resolved. Several articles by McCabe [10,11] are especially useful in examining the economics of serials publishing. While some economic incentives are important to stimulate the development of innovative applications of digital technologies to the publishing of scholarship, the terms for such services should not be allowed to impede the free flow of scholarly discussion on the global scale by placing barriers around the literature that inhibit access to it. The price academe pays for publishing services, regardless of who gets paid, should be reasonable and proportionate to the services actually rendered. We need to look for common ground with faculty and publishers in developing new ways to accomplish goals that are fundamental to a healthy scholarly communication system. These include rapid dissemination of research, quality control, maximum access, affordability, and maintenance of the scholarly record over time. If we do not find common ground with existing players in the system we need to cultivate relationships with new players.

INITIATIVES

Activity in the re-envisioning of scholarly communication is taking place on many fronts. In 1998 ARL formed the Scholarly Publishing and Academic Resources Coalition (SPARC).[12] SPARC quickly developed programs to actively create and support alternatives to high priced commercial publications for disseminating scholarship. Since then we have witnessed a sea change in the thinking of academe (faculty, scholarly societies, universities, and libraries), which is now reexamining past practices and thinking strategically about a future system of scholarly communication.

Since 1998 activity in this area is impressive. Some initiatives articulate conceptual frameworks and draft broad principles to guide development of a new global system.

One example is the Budapest Open Access Initiative,[13] formed in 2001, to launch an international effort to promote barrier-free dissemination on the Internet of scholarship in all disciplines. It is also working to make connections between separate, diverse open access initiatives across the world to promote innovation sharing, interoperability, and development of guidelines or best practices to promote the creation of a coherent, international system of scholarly communication.

The Zwolle Conferences,[14] held in 2001 and 2002, are another international effort focused on issues of copyright related to universities. The conferences drafted the Zwolle Principles that seek to establish a set of guidelines on copyright and intellectual property issues intended to promote core academic values including maximum access to research, academic freedom, and quality assurance mechanisms.

Lund University, in early 2003, announced its collaboration with the Open Society Institute and SPARC to create the Directory of Open Access Journals (DOAJ).[15] This project seeks to draw attention to open access journals, and to facilitate an effective cross-journal search system that will lay the foundation for a new system of scholarly communication.

The NEAR proposal,[16] the Tempe Principles,[17] and the Institutional Repositories initiative[18] are examples of efforts designed by library and academic communities to create alternative systems for scholarly communication and publishing.

Some initiatives are experimenting with new technologies, or practices, or disciplines: tackling specific or narrow technical and disciplinary issues. Pioneering examples of open access archives include Dspace[19] at MIT, eScholarship[20] for the University of California system, and Caltech's CODA.[21] The preprint server first modeled in 1991 by arXiv[22] in physics has proliferated to the extent that a Google search on "preprint servers" yields over 5,000 hits in 2003. Project Muse,[23] High-Wire Press,[24] and Project Euclid[25] are additional examples of library and academic initiatives that return publishing to academe. The article by Gass[26] provides a thoughtful discussion of several of these and other key projects.

What will the new system of scholarly communication look like? It's too early to tell. Ultimately, it is the needs of the faculty that must be supported since they develop and refine the content and put it to use in society. The fundamental purposes of scholarly communication (dissemination, review, organization, access and archiving) need to be retained in whatever system evolves. The challenge in the near future will be to pull the various initiatives and experiments together into a coherent whole that supports academic interests globally.

EFFECTING CHANGE

With so much activity on the national and international level what can an individual librarian contribute locally? The Create Change[27] material developed by ARL is a very helpful tool librarians can use to explain the roles, relationships, and issues surrounding scholarly communication to faculty and other colleagues.

Librarians can work to establish partnerships on campus between administrators, legal counsel, and other key stakeholders to examine the copyright transfer process. A service could be offered as a result of such a partnership to help interpret the impact of specific copyright transfer agreements, clearly identify future uses of the work the agreement authorizes, and suggest alternative language that faculty could insert to retain additional rights if necessary. Currently, at most institutions of higher education, faculty are expected to publish, and to do so independently from their home institution. While this independence is important it is also important to examine how the actions of one set of stakeholders impacts the others (e.g., copyright transfer affects journal access costs).

Librarians, in their capacity as liaisons to academic departments, can work individually with faculty to examine the copyright transfer agreements for the specific journals in which faculty publish or want to publish. Librarians can draw attention to the role copyright transfer agreements play in determining future access. Many copyright transfer agreements from publishers are changing to permit faculty to post their works on personal webpages and other digital venues. Yet many continue to require transfer that restricts access and legally compromises faculty rights to their own creations.[28] Appropriate copyright retention is important in reducing the amount of scholarship that is in effect privatized or "owned" by private commercial entities. When commercial publishers own exclusive copyrights they can set up pay-per-view and license terms that ensure their revenue stream and restrict rather than promote scholarly discourse. Librarians can also discuss these issues with their faculty who edit journals and can encourage editors to examine the pricing practices of the journals they edit. These activities would raise awareness of individual options related to copyright transfer and future access to scholarship.

We do not know exactly where digital technologies will take us, but retaining access rights within the academic community will be important. Librarians could launch a campaign:

"Copyright: it's yours. Keep it. You may need it."

ACCESS IMPACT

On the national or international level it would be useful to piggyback on work already in progress by Project RoMEO,[29] which is gathering copyright transfer agreements from key publishers and mounting them on the web for comparison. Project RoMEO is raising awareness internationally of the degree of rights the various agreements retain within academe. Building on that work librarians, or national library organizations could create an "Access Impact" factor, with tongue-in-cheek reference to ISI, to rate copyright transfer agreements. Faculty are already keenly attuned to Impact Factors and would readily understand this new terminology. Access Impact rating would help raise awareness of the legal and economic roles of copyright in the system as a whole. It would draw attention to the long-term implications of copyright transfer, its role in journal pricing and digital distribution options. It would also alert faculty to the moment at which they do hold the power to effect change. Access Impact ratings could build a clear and simple conceptual bridge between the concerns of librarians, faculty, and publishers.

CONCLUSION

We are moving from a known, mature system of scholarly communication in the print environment to an unknown digital environment in which many established practices are no longer deemed relevant or necessary. The stakes are high in changing the system both economically, for publisher profits and costs to higher education, and socially, for the potential benefits of open access to the scholarly record. Current experiments with technology and academic culture are disparate, creating models that one day may effectively be scaled up to rival what we have enjoyed in the print environment.

All this will take some time to unfold. In the meantime I believe that it is useful for librarians to look at the big picture of scholarly communication, to remember our contributions to the print environment, to think about what we can bring to the digital environment, and to be aware of the roles and stakes of faculty and publishers. It is also useful for us to strongly promote the values of librarianship, especially freedom of access, so that they are built into the fabric of the new system. It took the ingenuity and intelligence of our predecessors to create the past system and will require nothing less from us to create a new system of scholarly communication that fulfills its mission in the digital age.

REFERENCES

1. Joseph J. Branin and Mary Case. 1998. "Reforming Scholarly Publishing in the Sciences: A Librarian Perspective." *Notices of the AMS, 45,* 475-486. [cited February 18, 2003] Available at http://www.ams.org/notices/199804/199804-toc.html.

2. Jean-Claude Guédon. 2001. "In Oldenburg's Long Shadow: Librarians, Research Scientists, Publishers, and the Control of Scientific Publishing." Université de

Montréal. [cited February 18, 2003] Available at http://www.arl.org/arl/proceedings/138/guedon.html.

3. Pew Higher Education Roundtable. 1998. "To Publish And Perish." *Policy Perspectives*. Volume 7, Number 4. [cited February 18, 2003] Available at http://www.arl.org/scomm/pew/pewrept.html.

4. Association of Research Libraries. 1998. "ARL/AAU Pew Roundtable on Managing Intellectual Property in Higher Education." [cited February 18, 2003] Available at http://www.arl.org/scomm/pew/.

5. Anthony M. Cummings et al. 1992. "University Libraries and Scholarly Communication: Study prepared for the Andrew W. Mellon Foundation." The Association of Research Libraries, Washington, DC.

6. Association of Research Libraries. "ARL: A Bimonthly Report on Research Library Issues and Actions from ARL, CNI, and SPARC." Scholarly Communication link. [cited February 18, 2003] Available at http://www.arl.org/newsltr/osc.html.

7. Peter Suber. 2003. "Free Online Scholarship Newsletter: How the Internet is Transforming Scholarly Research and Publication." [cited February 19, 2003] Available at www.earlham.edu/~peters/fos.

8. Pew Higher Education Roundtable. 2.

9. Siva Vaidhyanthan. 2001. *"Copyrights and Copywrongs: The Rise Of Intellectual Property And How It Threatens Creativity."* New York and London. New York University Press.

10. Mark McCabe. 1998. "The Impact of Publisher Mergers on Journal Prices: A Preliminary Report." ARL, the Newsletter of the Association of Research Libraries. [cited February 19, 2003] Available at http://www.arl.org/newsltr/200/mccabe.html.

11. Mark McCabe. 2001. "The Impact of Publisher Mergers on Journal Prices: Theory and Evidence." *The Serials Librarian*. Volume 40, Issue 1/2.

12. Scholarly Communication and Academic Resources Coalition–SPARC. [cited February 18, 2003] Available at http://www.arl.org/sparc/index.html.

13. Open Society Institute. 2001. "Budapest Open Access Initiative (BOAI)." [cited February 18, 2003] Available at http://www.soros.org/openaccess/.

14. SURF Foundation. 2002. 2nd Zwolle Conference. "Copyright and Universities: From Principles to Practices." [cited February 18, 2003] Available at http://www.surf.nl/copyright/zwolle/.

15. Lund University, Open Society Institute, SPARC. 2003. "Directory of Open Access Journals." [cited February 19, 2003] Available at http://www.doaj.org/.

16. Shulenburger, David E. 1999. "Moving with Dispatch to Resolve the Scholarly Communication Crisis: From Here to NEAR." *ARL: A Bimonthly Newsletter of Research Library Issues and Actions*. Number 202. [cited February 19, 2003] Available at http://www.arl.org/newsltr/202/shulenburger.html.

17. Association of Research Libraries. 2000. "Principles for Emerging Systems of Scholarly Publishing." (The Tempe Principles) [cited February 19, 2003] Available at http://www.arl.org/scomm/tempe.html.

18. Raym Crow. 2002. "The Case for Institutional Repositories: A SPARC Position Paper." ARL Bimonthly Report 223. [cited February 19, 2003] Available at http://www.arl.org/newsltr/223/instrepo.html.

19. Massachussetts Institute of Technology. "Dspace." [cited February 19, 2003] Available at http://libraries.mit.edu/dspace.

20. University of California, California Digital Library. "eScholarship." [cited February 19, 2003] Available at http://repositories.cdlib.org/escholarship/.

21. Caltech Library System. "Caltech CODA." [cited February 19, 2003] Available at http://library.caltech.edu/digital/.

22. Cornell University. "arXiv.org." [cited February 19, 2003] Available at http://arXiv.org/.

23. Johns Hopkins University Press and Milton S. Eisenhower Library. "Project Muse." [cited February 19, 2003] Available at http://muse.jhu.edu/.

24. Stanford University Library. "High Wire Press." [cited February 19, 2003] Available at http://highwire.stanford.edu/.

25. Cornell University Library. "Project Euclid." [cited February 19, 2003] Available at http://projecteuclid.org/Dienst/UI/1.0/Home.

26. Steven Gass. 2001. "Transforming Scientific Communication for the 21st Century." *Science & Technology Libraries.* Volume 19, Number 3/4.

27. Association of Research Libraries, Association of College and Research Libraries, and SPARC. "Create Change: A Resource for Faculty and Librarian Action to Reclaim Scholarly Communication." [cited February 19, 2003] Available at http://www.createchange.org/home.html.

28. Joint Information Systems Committee. "Project RoMEO (Rights MEtadata for Open Archiving)." [cited February 19, 2003] Available at http://www.lboro.ac.uk/departments/ls/disresearch/romeo/.

29. Project RoMEO.

DIGITAL ARCHIVE AND RETRIEVAL

Issues and Concerns
with the Archiving of Electronic Journals

Janet A. Hughes

SUMMARY. The advent of electronic scholarship brings new opportunities to authorship, publishing and librarianship. However, like all new paradigms, it brings concomitant challenges, including how to maintain access to electronic information through time, also known as digital archiving or electronic archiving. Electronic archiving can involve the archiving of discrete items, such as books, that have been digitized, the archiving of items that were "born digital" with no print counterparts, and the archiving of serial publications that have both print and electronic formats, which are not always equivalent. This paper discusses the policies,

Janet A. Hughes, MLIS, is Biological Sciences Librarian, The Pennsylvania State University, 408 Paterno, University Park, PA USA 16802-1811 (E-mail: jah19@ psu.edu).

[Haworth co-indexing entry note]: "Issues and Concerns with the Archiving of Electronic Journals." Hughes, Janet A. Co-published simultaneously in *Science & Technology Libraries* (The Haworth Information Press, an imprint of The Haworth Press, Inc.) Vol. 22, No. 3/4, 2002, pp. 113-136; and: *Scholarly Communication in Science and Engineering Research in Higher Education* (ed: Wei Wei) The Haworth Information Press, an imprint of The Haworth Press, Inc., 2002, pp. 113-136. Single or multiple copies of this article are available for a fee from The Haworth Document Delivery Service [1-800-HAWORTH, 9:00 a.m. - 5:00 p.m. (EST). E-mail address: docdelivery@haworthpress.com].

10.1300/J122v22n03_10

problems, questions, and concerns of archiving serial publications that have both print and electronic versions. *[Article copies available for a fee from The Haworth Document Delivery Service: 1-800-HAWORTH. E-mail address: <docdelivery@haworthpress.com> Website: <http://www.HaworthPress.com> © 2002 by The Haworth Press, Inc. All rights reserved.]*

KEYWORDS. Electronic journals, digital archiving, conservation and restoration

INTRODUCTION

Even as print journal subscriptions are cancelled in favor of electronic versions, and as digital collections become more prevalent, librarians are questioning the stability, reliability and future of the electronic resources upon which they have become reliant. What assurances are there that the electronic resources librarians are paying for today will be available to serve future scholars? Are publishers and information providers planning for the long term? What policies, processes, and economic models will mold future access? This article is a summary of the concerns, questions, policies, processes, problems and solutions that were found when investigating the current status of, and research on, archiving of electronic resources.

BACKGROUND

The advent of electronic scholarship brings new opportunities to authorship, publishing and librarianship. However, like all new paradigms, it brings concomitant challenges, including issues of how to cite electronically available articles that do not follow the standard volume/issue format, how to allow services like interlibrary loan while preventing copyright infringement, how to establish first publication dates and canonical versions, and how to maintain access to electronic information through time, also known as digital archiving or electronic archiving.

There are three main areas of concern in electronic archiving, the archiving of discrete items, such as books or photographs, that have been digitized, the archiving of items that were "born digital," that is, items that have no print counterparts, and the archiving of serial publications that have both print and electronic formats, which are not always equivalent. Although "born digital" archiving will become more important as electronic scholarship moves toward that paradigm, at the moment, archiving of serial publications that have both print and electronic formats is a more pressing issue. This paper focuses on that kind of archiving.

Journals serve two main functions, to transmit information through space, and to transmit information through time (Neavill and Sheblé 1995). Electronic versions of journals perform the first function admirably, but they raise questions about the latter.

Archiving of print materials has been the domain of libraries for many years. Archiving print items requires only that the items are kept intact in some suitable long-term storage facility. As long as the contents are in a language that is still understood, and the items can still be physically handled, no other equipment or accessories are required. However, with electronic archiving, there are added problems because it is not enough to merely preserve the physical presence of the items–they must also be accessible, verifiable, and usable. That means not only must the contents be available, they must be easily accessible, they must be verified to be uncorrupted, and they must be in a format that can be used by the currently available equipment and software. These requirements add a whole additional layer of technical problems to the equation.

In addition to technical issues, there are policy issues that were not relevant with print subscriptions where individual libraries purchased print journals and made local decisions about retention and archiving. Online journal subscriptions are more akin to leasing information, like the access fees paid for online indexing and abstracting services. With such online databases, annual fees for access were paid, with the understanding that if a subscription were ever cancelled, all access was relinquished. However, most librarians do not find this scenario acceptable with online journal subscriptions. While some progress has been made in resolving this dichotomy between librarians' expectations and the licensing agreements for most current electronic journals, there is very little consensus on how this should work. Many libraries are still uneasy about electronic archiving, choosing to also retain print subscriptions (Flecker 2001; Reich 2002). However, as serials budgets get tighter, format duplication in journals cannot be supported (George 2002; Gyeszly 2001; Hunter 2001). Furthermore, many libraries are switching over to electronic format only, because users seem to prefer the convenience of online access to journals over print access, to save money and shelf space, and because there is a sense that you get more added value with online journals (Butler 1999; Goodman 2001; Hunter 2001; Kenney and McGovern 2001; Malinconico 1996; Montgomery and

King 2002). Therefore, this policy issue will only become more relevant as more libraries follow this trend.

There are a number of issues concerning the use of electronic versions for interlibrary loan and other sharing agreements. Copyright holders are concerned about the possibilities of infringement with digital information, because copying and distributing are much easier with electronic journals than with print (Davidson 2000; Douglas 2000; Hunter 2001; Morris 2000). Therefore, control over electronic information, including electronic archives, becomes another issue that was never a problem with print archives and fair use doctrine.

And, of course, there is the issue of money, that is, who should pay for electronic archiving, and how should that payment be structured? Unlike print subscriptions, where the pricing for journals was fairly standard for most institutions in a region, regardless of size, most publishers seem to think the number of persons served by electronic resources, even archival resources, should be included in pricing structures.

These issues and others create the quagmire that is digital archiving.

METHODS

To find out what major publishers were doing about electronic archiving of their serial publications, several publishers' web pages were searched to find their policies and ideas on archiving. The results of the search were disappointing. Policies could not be found on many sites, some policies found were so brief as to be completely non-illuminating, and only a few policies were easily located, coherent, and comprehensive (see Table 1). To augment the paucity of information a brief survey concerning archiving policies was sent to several large publishers (see Appendix). Although some representatives forwarded the questions to the appropriate authorities, only one publisher responded with a somewhat sheepish admission of a lack of a true policy.

Next, a literature search was done to discover what librarians and others were doing and thinking about electronic archiving. As suspected, many others have long been discussing the issues. However, there seem to be many, many questions and concerns, and relatively few answers in the literature, although some pilot projects look promising. Questions abound, such as, who is best suited to do archiving, the publisher or copyright holder, libraries, or some' third party? How long should archiving last? Is archiving permanent, as is the assumed case for most print items, or should archiving be only for as long as the infor-

TABLE 1. Publishers' Electronic Archiving Policies

Publisher	URL	Comments
American Chemical Society	http://pubs.acs.org/archives/index.html	FAQ is reasonably comprehensive but difficult to piece together
American Institute of Physics	http://www.aip.org/journals/archive/archive.pdf	Brief but covers some issues, including data refreshing, lapsed subscriptions
American Mathematical Society	http://www.ams.org/jourhtml/about.html	Short paragraph guaranteeing perpetual archives; deposit at JSTOR; offers archival disk to subscribers
BioMed Central	http://www.biomedcentral.com/info/about/	No policy listed; articles are archived by PubMed Central; supports multiple digital repositories
BioOne	http://www.bioone.org/bioone/?request=get-help-faq	Short paragraph allows access after cancellation; perpetual content archiving
Blackwell	http://www.blackwellpublishing.com/jnl_default.asp	Policy not found; no survey reply
Cambridge	http://journals.cambridge.org	Policy not found; FAQ briefly mentions continued access after cancellation
Elsevier (includes Academic IDEAL)	http://www.elsevier.com/inca/publications/misc/ni2164.pdf	Comprehensive policy easily found; answers most outstanding concerns
Highwire	http://highwire.stanford.edu/about/intro.dtl	Could not find policy; many societies and publishers—does each have its own policy? free access after embargoes
Kluwer Online	http://www.lwwonline.com/common/kap/gw-faq.htm	Short FAQ paragraph guarantees access through site for 4 years; local archiving
Project MUSE	http://muse.jhu.edu/proj_descrip/back_iss.html	Local archiving because institutions own the data; promises of perpetual archiving
PubMed Central	http://www.pubmedcentral.com/about/intro.html	Policy not found, only a mention of the fact that archiving is part of its mission
Royal Society of Chemistry	http://www.rsc.org/is/journals/archive.htm	Short paragraph guarantees perpetual access, even after cancellation
SPARC	http://www.arl.org/sparc/core/index.asp?page=a0	Simple paragraph advocating comprehensive archiving
Wiley Interscience	http://www3.interscience.wiley.com/about.html	Policy not found; no survey reply

117

mation contained therein is considered useful? Who makes that judgment, and what criteria do they use to make it? What should be archived? Should everything be archived, or just the main articles, and again, who makes that judgment? Should the archive be easily and instantly accessible, or more a repository, as print archives are? Should archives be centralized or local? How much redundancy is required? What technical solutions exist, and should multiple techniques be used to ensure that at least one remains viable? What will access policies include? With print subscriptions, canceling the print still allowed retention of the purchased information, but what of online subscriptions– what happens if a subscription is cancelled? If the journal was bought as part of a package, can it be cancelled at all? If cancelling a subscription is allowed, does the contract allow cancelling only the print and still paying to retain online access? If all access is lost by cancelling, are there ways to compensate? What happens with an active subscription if the publisher decides to no longer maintain the backfiles? Should access rights be included in licenses? There has been some movement to include perpetual access in licenses, but this is not yet the norm (Graham 2000). Also, should libraries expect to retain access rights after a journal changes publishers, or ceases to exist?

All these questions seem to boil down to several main issues discussed in the literature about archiving: (1) what is meant by electronic archiving, including what to archive and how long to archive, (2) who should do the archiving, (3) how should archiving be done, including all the technological issues, (4) how can integrity be protected, (5) how should access to archives be priced, and (6) what should the policies be for access after cancellations or changes in publishers.

SUMMARY OF LITERATURE

What Does Electronic Archiving Mean?

An archive "is a collection of material which someone has vowed to maintain for the foreseeable future and that offers some kind of access to the collection" (Feldman 1997, page 52). While that may seem a vague definition, it allows the leeway needed when approaching such a nebulous subject. The main issues involved when describing what is meant by the term electronic archiving include discussions of accessibility, location, content, and longevity.

Several years ago, much of the discussion was about the accessibility of electronic archives, although this issue seems less frequently mentioned in current literature. The question was whether electronic archives should be "open," that is, instantly accessible, or could they be "dark," acting as repositories only. The advantages of "open" archives include the instant gratification of access, the idea that accessible archives would be used more, and therefore would have greater value, and the thought that information that is regularly used might be less likely to become technically obsolete. The advantages of "dark" archives would be that simple repositories would not pose competition for publishers who may want to market their own archives, and it might also be relatively cheaper to maintain repositories since these would require neither a current interface, nor the cost of administering access rights (Flecker 2001). However, the probability increased that the undisturbed, unused information may become obsolete. Although some literature still mentions the possibility of "dark" archives (Anonymous 2001; Flecker 2001; Kenney and McGovern 2001), most recent literature seems to assume that archives will be "open." Most publishers' policies that were found seem to imply "open" archives for current subscribers. However, it still has not been decided whether the data may be segregated into deep backfiles that are a little less accessible versus a more current back file. As archives become larger and more time-consuming to search and manage, this may become more of an issue.

Location of archives is an issue that seems to be finding some solutions. The question involved is whether archives should be centralized or scattered. This answer depends very much on who does the archiving. If it were assumed that large publishers, consortia or national institutions would do the archiving, obviously these would be centralized. If instead it were assumed that individual libraries would archive, the archives would be distributed. There are advantages to both scenarios. A central stable archive would present a single interface, and allow search and retrieval across journals from various subjects, and if done by a national institution or consortium, materials could come from various publishers. The latter is especially appealing because users generally want information by subject, not by publisher (Rowe 2002; White 2001). Also, it may be more cost effective to have centralized archives, because the costs of upgrading equipment and migrating data would have the advantage of scale. However, unless that centralized archive has considerable redundancy built into its infrastructure, there could be access problems when a router goes down or there is some disaster at the centralized site. Also, the end users of the electronic archives might have very little control over the content of the archives. Scattered archives allow local control over the archives, and the redundancy involved ensures that at least one copy is always accessible. The LOCKSS (Lots of Copies Keep Stuff Safe) project advocates a scattered, local archiving approach where libraries purchase, rather than lease, the electronic information, and are given rights to make local caches that act as backups to the publishers' sites. Such a system allows local control over what is archived, and the numerous local caches allow redundancy and error checking (Reich 2002; Reich and Rosenthal 2001). One possible solution is a combination of centralized and localized archives. Elsevier policy, found at http://www.elsevier.com/inca/publications/misc/ni2164.pdf, allows for both centralized and distributed archives. The company itself will archive in several places, and its policy allows its Science Direct OnSite customers to archive locally. Furthermore, the company is working with other institutions to deposit archives in neutral sites (Hunter 2001). Project MUSE (http://muse.jhu.edu/proj_descrip/back_iss.html) and Kluwer Online (http://www.lwwonline.com/common/kap/gw-faq.htm) briefly mention allowances for local archiving and promises to centrally archive.

The content of archives poses a huge problem. What should be archived? Should archiving include everything or only selected items? Librarians need to discriminate on what to buy and what to keep with print materials, due to cost and space constraints. Should this philosophy apply to electronic information also? Many authors seem to believe so, advocating that it is necessary to establish criteria to determine which content should be preserved (Keyhani 1998). Some suggest keeping an item only if it is of generally high quality or from a reputable source, or if it is unique, or beautiful or useful (Gilson 1997; Feldman 1997). Some would argue that everything should be kept, and can be weeded out later (Feldman 1997). It probably should depend on the archiving institutions' missions and philosophies–is archiving done for research, for cultural heritage, for legal reasons, or for consumer interest? If archiving is done for research purposes, information in peer-reviewed journals may suffice (Day 1998). However, if peer-reviewed journals are kept, what cultural information will be lost when the newsletters, bulletins, and other grey literature are no longer accessible? Also, even within peer-reviewed journals, should all the bits be archived? Are advertisements important, and what of job notices, conference announcements, obituaries, and letters to the editor? What about items like the listing of the editorial boards? Many online journals have such a page, but these tend to be updated whenever there are changes, obliterating the older information (Arms 2000). LOCKSS advocates having this kind of information online so that the "snapshot" caches would reflect this information (Reich 2002). It is difficult to make decisions about the relevance and importance of "non-research" information, because even items now considered trivial may be important to future historians to understand this stage in history (Morris 2000). A further problem is that the standards for tagging non-article content are not strong. For example, Elsevier does not provide full-text standard generalized markup language (SGML) files for "other content" without research significance, and portable data format (PDF) files and SGML files are not done for items not listed on the tables of contents, precluding cover-to-cover archiving (Inera Incorporated 2001). This issue may be moot, though, because even now, the advertisements in such journals as *Nature* are not found in the online versions. Therefore, the likelihood of such ephemeral materials being electronically archived is slim.

Even after the choices have been made about what items to archive, there is the conundrum of links and linked resources. Is it important that all links in an archived item be kept, and that all the resources to which there are links be archived also? What about the links in those linked items? Where should the boundary be made, and if no boundary is established, would that be akin to archiving the entire Internet? Can such an undertaking be even contemplated? Some authors have discussed possible mechanisms for archiving the Internet (Feldman 1997; Masanes 2002). However, this is not yet a widely accepted vision, and some limits should be established. Most publishers' policies do not directly address such boundaries. This leads to the important but unresolved issue of what is the version of record for a journal, the print or the online version, and if it is the online version, which edition of the online version is the definitive one (Mabe 2001)? Some publishers insist the print version is still the canonical version, despite differences in content and organization (Terry 2001). Others argue that paper versions are not the real journals anymore since online versions offer so many other resources, like video, machine-readable tables of data, audio clips, etc. (Boyce 2000; George 2002; Hunter 2000; Morris 2000; Tenopir 1999). Journal articles no longer exist as discrete entities; they work with and may be dependent on other journals, abstract databases, and other sources of electronic information (Boyce 2000). If all the links and the resources to which the links lead are considered to be part of the canonical version, these linked resources must be archived, or the usability of the online resource may be compromised. However, there are several problems with this idea of comprehensive archiving. There may be copyright issues since some of the linked items may come from other publishers, personal web sites and other places with which the archiving institution does not have agreement. Links may also lead to password restricted or required registration sites, or to moved pages and broken links (Feldman 1997). Also, added materials may use proprietary software that would be impossible to keep updated without ongoing agreements with the software companies. There would have to be collaboration between all interested parties in order for this vision of archiving to proceed (Boyce 2000; Feldman 1997; Gilson 1997). Furthermore, although costs for the actual storage of all such resources may decline, the costs of administering such massive collections will surely rise. There are so many obstacles to comprehensive archiving that it is rarely discussed as a feasible option. One of the more promising current projects being tested will not allow for comprehensive archiving like this. The LOCKSS project makes "snapshot" caches as information is put online, but does not include dynamic materials such as java scripts, links to video, etc. (Reich 2002; Reich and Rosenthal 2001). Thus, this issue of how much of the "other resources" to electronically archive is still unresolved.

Longevity of archives is another contentious issue. How long should items be archived–permanently or for some set time? Most of the literature assumes long term (Flecker 2001; Hunter 2001). Some mention a nebulous period of the length of time that such information retains value for its users (Gilson 1997). Previously, it was suggested that scientific and technical information would only have a short "shelf life" and so ten years might be sufficient (Tagler 1998). However the American Chemical Society (ACS) and the American Physical Society (APS) think there is a market for older materials, and are digitizing their complete backfiles. The feasibility study done for Harvard assumes that the archive would remain for at least 75 years (Inera Incorporated 2001). There does not seem to be any accord for the time period that electronic archives should exist. Furthermore, there are no established criteria for making decisions about retention. Some authors suggest the community of users of archived items are the only ones suited to making retention decisions, because only they know what is valuable to them (Rowe 2002). However, if the users are not the archiving bodies, will they have any input? If left up to archiving bodies, which may be commercial ventures, will economic factors trump research value? Whatever the criteria for retention at individual archives, if online subscriptions are to be treated like print subscriptions, which seems to be the attitude of many librarians (Buckley et al. 1999), then at least a few copies must be permanent. Whether these permanent copies are fully functional and easily accessible has still not been resolved.

Thus, the question of what is meant by electronic archiving is still very much under debate. Although the consensus seems to be for accessible archives, there is still some merit in "dark" archives. Both centralized and scattered archives have proponents, but most literature seems to favor a combination of both. The content of archives is completely uncertain, with most advocating some discrimination, but no agreement exists on the boundaries around archived items. And finally, although many assume permanent archives, there is considerable debate about the feasibility of such permanence.

Who Should Do Archiving?

There are several possible entities to do the archiving: publishers, individual libraries, and consortia/national institutions/third party sources. There are many advantages and disadvantages to each of these possible archiving bodies.

Many publishers, especially large publishers, are proceeding as if they would be the archiving source. This allows them to receive any financial benefits archives might bring, but more importantly, the publishers would retain control over the information. The issue of control is very important to publishers, but it acts as a barrier to access (Davidson 2000; Douglas 2000; Morris 2000). Publishers acting as archival institutions leave libraries without any control over the contents of archives, and at the mercy of those publishers with respect to access and pricing. Commercial publishers must balance the costs of keeping control over their information, with the diminishing income that such archives might bring, especially in scientific and technical information where the use of older materials is infrequent. Thus, many authors only trust publishers to retain archives as long as those archives are profitable (Anderson 2000; Douglas 2000; Morris 2000; Rogers 1999). Such authors emphasize that archiving is not a core competence of publishers, and publishers' goals and business models do not usually intersect well with libraries' goals (Anderson 2000; Kenney and McGovern 2001; Keyhani 1998; Mabe 2001). Furthermore, although large publishers, such as Elsevier and ACS, are big enough that they could probably be trusted to remain in business and thus keep the archives intact, small publishers, that could cease operations or be bought out by larger corporations, should not be entrusted to maintain archives (Arms 2000). Even if publishers can be trusted to archive, they must cooperate with the other stakeholders, such as libraries, to ensure long-term access and preservation of their online serials (Arms 2000; Day 1998). Several current projects, funded through the Mellon/CLIR (Council on Library and Information Resources) initiative, are investigating the feasibility of large universities and institutions working with specific publishers to archive electronic journals (Albanese 2001; Flecker 2001; Hunter 2001).

One minor aside is the idea of self-publishing and self-archiving. If authors retain copyright ownership, self-archiving is possible (Butler 1999; Harnad 2001). However, refereeing is still important in establishing the credibility of articles. This self-archiving idea would be limited to those disciplines, such as physics, where it is acceptable to have electronic pre-prints and still have those articles published later in refereed journals. It would not address the issue of archiving the final published articles. Furthermore, few authors have the time, expertise or inclination to do electronic archiving themselves, and would require help from their institutions and libraries. Thus, although this kind of archiving has been discussed frequently in the scientific literature, it does not truly address the problem of electronic archiving of journals.

Many people assume that libraries will do the archiving because preserving information is a traditional mission of libraries (Buckley et al. 1999; Luijendijk 1996; Neavill and Sheblé 1995). In order for individual libraries to do archiving, there must be technical expertise, equipment, funding, and permission to make local caches of the information. Such permission was rarely included in electronic licenses in the past, although progress is being made in that area to allow libraries to actually own the information purchased. The LOCKSS project works on that premise (Reich 2002; Reich and Rosenthal 2001). LOCKSS does not require much funding or technical expertise, and thus it shows much promise. However, the caches created in LOCKSS do not archive the linked resources; the caches are akin to "snapshots." Thus, this project will not work for all electronic journals, and the LOCKSS itself endorses using several solutions to ensure archiving. Other individual archiving solutions seem to require more funding than most libraries can spare. Thus, some authors suggest collaborative efforts between libraries to create financially feasible archives (Keyhani 1998).

This leads to the idea that large consortia, national institutions or other third party entities might be best suited for the job of archiving (Arms 2000; Buckley et al. 1999; Gilson 1997). Such large institutions would have the stability for such an undertaking. Furthermore, because they would not necessarily be limited to a single publisher's content, subject access would be enhanced. Knowledge should be preserved and organized by discipline rather than by publisher, because users want information that way (Anonymous 2001; Rowe 2002; White 2001). However, the funding still may be problematic, especially for publicly funded institutions, and these large institutions would still need to get permission to archive the information. Thus many parties assume that there should be third party archiving independent of publishers but with active partnerships with publishers (Anderson 2000; Boyce 2000; Day 1998; Flecker 2001; Hunter 2001; Reich 2002; Reich and Rosenthal 2001). JSTOR is often cited as an example, even though its high fees may present a barrier to access (Anderson 2000; Chepesiuk 2000; Davidson 2000). Also, there have been some suggestions that JSTOR has shied away from much of the scientific and technical material due to the vast quantities of information available, the costs of doing optical character recognition with scientific formulae and symbols, and the limited shelf-life and therefore limited commercial return of scientific and technical journals (Tagler 1998). Several publishers, such as Elsevier, Kluwer, and the Royal Society of Chemistry (http://www.rsc.org/is/journals/archive.htm), briefly mention depositing archives with third party sources in their policies but usually only as a last resort. One third party option that was intriguing was the idea that there might be a new kind of business created, similar to print journal back issue dealers, who would archive backfiles of electronic journals (Luijendijk 1996).

Assuming some level of third party archiving, a team at the Library of Congress is investigating a four-tier architecture for archiving. The lowest tier would be little more than a repository for bits, a data warehouse managed by some trusted third party. The next tier would be a gateway that would control how the bits are accessed by knowing the address of particular bits and returning a set of bits when requested by a tier three collection. Tier three collections would be different institutions that would produce metadata associated with the data to provide context to the bits, such as whether these are text or images, and what software is needed to render the bits properly. Such collections would also manage the restrictions and terms for access, and provide the fourth tier, an interface level at which the patrons access the information (Friedlander 2002).

No matter who does the archiving, although many authors seem to be leaning towards some third party solution, there will be issues and concerns for all the stakeholders. The best solution is probably some combination of all the above. In any case, it has been suggested that a coordinating body may be required since materials suitable for archiving come from many sources whose priorities differ, making it unlikely that a common archiving policy can be created in the near future (Feldman 1997; Haynes and Streatfield 1998).

How Should Electronic Content Be Archived?

The technical problems with electronic archiving are many, but the issue that receives the most prominence is the rapid obsolescence of the hardware and software on which the journals are dependent (Day 1998; Feldman 1997; Mabe 2001; Neavill and Sheblé 1995). It is irrelevant to have long-term storage media if the hardware needed to use these becomes obsolete (Day 1998). Similarly, it is useless to maintain hardware if the software used to access the materials becomes degraded. Thus, electronic information must be kept refreshed, and software and hardware used to access it must be kept current. How to keep the data refreshed and the hardware and software from becoming obsolete is a major problem that has several possible solutions.

There are several levels of electronic archiving or preservation possible: conservation, preservation of access and preservation of content. Conservation retains the full look and feel of the publication, preservation of access requires maintaining the indexes, metadata, etc., beyond just the articles, and preservation of content only is akin to data warehousing (Arms 2000). Preservation of content only is generally not considered acceptable. Thus, any method used to archive electronic journals must either conserve the look and feel of the journal, a strategy known as emulation, or it must consider content, access and functionality of the journal paramount, allowing for data migration. The choice between these two strategies has not been fully resolved, although migration is most often mentioned in the current literature.

Items that should be emulated are ones where the look and feel are important, such as articles which include biotechnological computer-enabled experiments, visualizations and games (Friedlander 2002). Emulation is the most difficult strategy (Arms 2000), and would require preserving the content and enough metadata to adequately describe systems so that future emulators could mimic the behavior of obsolete hardware and software (Day 1998; Neavill and Sheblé 1995). While there are enough aficionados of old video games who create emulators of Pac-Man and Pong and post them freely on the Internet, it is unlikely that there will be similar fans of electronic journals willing to create emulators for free. Thus, emulation will probably require expensive technical expertise in order to succeed.

Migration is the strategy most often cited as having the best potential, and most of the publishers that mentioned refreshing of data implied a migration strategy. Items where the content and access are most important can be migrated. Migration involves progressively moving the content to current software and hardware, while acknowledging that migration changes content, and it may not always be possible to preserve every aspect of the original (Morris 2000). Migration strategies used should be recorded as metadata and preserved together with the original item so that future users are aware of significant changes made to a document during the preservation process (Day 1998).

There are two file formats used in electronic journals that are amenable to archiving, SGML and PDF formats. Using SGML as the underlying language seems to lend itself to migration. SGML is the current standard (Boyce 2000). SGML retains the structural information about the content of a document, and tagging allows linking and indexing and searching now and in the future. Also, it is human readable and non-proprietary, so likely to be usable in the future. Also, it can be converted to XML, a subset of SGML, which is touted as the standard of the future (Inera Incorporated 2001). Boyce envisions that publishers could make the electronic archival quality master copy in SGML first, and create the published version from that. Migration and updates would be done on the master copy by automatic computer scripting and programs to maintain technological currency (Boyce 1998). Problems may still exist, however, because publishers are still using proprietary add-ins, and other publisher specific processes, which will make migration more difficult. For example, at the moment Elsevier uses SGML for its articles, but it keeps all formatting separate on style sheets that are not archived, making it difficult to generate the same look after migration (Inera Incorporated 2001). Similarly, the Association for Computing Machinery (ACM) uses SGML for its digital library, but the document type definition (DTD) is proprietary, and the special algorithms used to render mathematical symbols and equations from SGML are specific to ACM (Arms 2000). On the other hand, the University of Chicago keeps the boilerplate text and face markup in the SGML file so that the SGML accurately preserves text as published (Inera Incorporated 2001).

If the goal of archiving is to replicate the print version accurately, such as with JSTOR, PDF format works best (Anonymous 1999; Chepesiuk 2000). Many publishers are using PDF files as the de facto standard (Buckley et al. 1999). PDF retains the visual presentation of the original but not the links, or the linked resources, and it is not searchable (Anonymous 1999). If those links and linked resources are considered important, as discussed above, PDF files are not sufficient (Boyce 2000). Also, because the PDF format is proprietary, its continued availability is unknown. Many publishers, such as Elsevier, are thus keeping both SGML files and PDF files to allow for full usability and visual presentation.

Whatever file format is used, standards are needed to ensure interoperability among preservation schemes (Feldman 1997; Flecker 2001; Solla 2002). Normalizing content would reduce complexity, although it may cause information loss (Flecker 2001). The study on electronic archiving feasibility looked at making a standardized DTD to preserve the intellectual content of journals independent of the format in which the content was originally delivered. Thus it would keep the data but not the look and feel of the journals (Inera Incorporated 2001). Such normalization has advantages, but persuading publishers to change their proprietary DTDs may be difficult.

The question of how to archive is still fraught with many problems. The consensus seems to be to use SGML and PDF as file formats, and to migrate data when necessary. However, without standards none of this will be easily accomplished.

How Can the Authenticity and Reliability of Electronic Archives Be Assured?

Electronic journals are malleable, which allows updating and revisions. However, there is no accord on the degree of alterations that can or should be allowed. Can simple errata be inserted, or should nothing be changed? (Buckley et al. 1999). Preprint archives allow revisions. The abstracted reviews intended for *Reference Reviews Europe Annual* are posted online prior to publication, and after the print publication is released, the online reviews are edited and revised (Terry 2001). Among publishers, only the American Institute of Physics (http://www. aip.org/journals/archive/archive.pdf) clearly states that original content would never be altered, but would be annotated and supplemented by clearly noted errata and references. This lack of consensus leads to the problem of reliability and authenticity. Users want to be assured that the items they read accurately reflect the original. It is difficult to prove that digital information has not been corrupted over time and there is a great deal of skepticism that authenticity can be guaranteed over the long term (Davidson 2000).

There are several ways to ensure authenticity. Using PDF files that are static makes them relatively difficult to change and therefore these would have reliability. However, PDF files reduce the functionality of electronic archives. It might be necessary to have both a static form, like a PDF file, plus a dynamic version, like SGML, the former to preserve the original look and feel as well as provide reliability, and the latter to provide the functionality. Another way to ensure reliability is through redundancy. LOCKSS uses peer-to-peer comparisons among many local caches to check for damage and changes (Reich 2002; Reich and Rosenthal 2001).

There can be authentication and identification of successive versions through digital watermarks. These include some encryption systems that have public and private keys to encrypt and decrypt, hashing, which are digital fingerprints, and time stamping, which takes hashing algorithms and verifies the time and date of creation in relation to other documents (Neavill and Sheblé 1995). This way users understand that there may have been updates and revisions by knowing the version they are viewing (Day 1998).

More than likely, a combination of digital watermarks, redundancy, and static formats will be required to diminish users' mistrust of the authenticity of digital archives.

What Pricing Models Will Work for Electronic Archiving?

There is very little in the literature about the pricing of archives. For the moment, the pricing of most electronic journals is based upon print subscription prices (Hunter 2001). However, unlike print subscriptions, the number of potential users is often used to calculate the price, causing larger institutions to pay much more than smaller ones. Pricing will depend on who does the archiving. If publishers do the archiving themselves, they will need to price access in such a way as to make sense financially (Douglas 2000). Elsevier includes the cost of access to the archives in their Science Direct fees. However, they are also making backfiles available for a one-time extra payment. ACS Web Edition subscription fees only give access to the current year plus a rolling back file of four years. Access to earlier years in the ACS Journal Archives requires an additional annual fee. If individual libraries do archiving using the LOCKSS system, the minimal costs are carried by the individual libraries, which are responsible for the purchase of their subscriptions, their computer equipment, and the maintenance of the system. The software itself and participation in the project are free, but there is a voluntary fee service, the LOCKSS Alliance, which will provide technology support and collection coordination services (Reich 2002). If public institutions undertake electronic archiving, profitability is moot, although cost recovery might be desirable. Continual funding through taxes, fees, or some other arrangement, will be essential (Davidson 2000; Flecker 2001; Gilson 1997; Morris 2000). Other third party archiving solutions, such as PubMed Central, have yet to show that providing free access to backfiles is a viable business model. JSTOR, the third party solution most often noted, is a non-profit organization, but its access fees are considered high (Chepesiuk 2000).

Pricing will continue to be a contentious issue as more archives are created. At the moment, there are several models of pricing, and this issue can only be resolved after the issues of who does the archiving, how archiving should be done, and other aforementioned issues are resolved.

What Models Exist for Continued Access?

Most librarians want to treat electronic subscriptions the same as print subscriptions, and expect to retain access to the information purchased even after cancellation of a subscription (Buckley et al. 1999). More licenses are beginning to include some provision for access for lapsed subscribers (Goodman 2001), but this is not yet the norm (Graham 2000). Publishers are aware of the desires of librarians, but have chosen many different approaches to provide perpetual access. ACS Web Editions current subscription fees only include a rolling back file. Subscribers must pay extra for access to the archive, and lose access to the archive if the annual fee is not paid (see http://pubs.acs.org/archives/index.html). Elsevier allows access to backfiles if a subscription is cancelled but there still is an active Science Direct subscription. If all access to Science Direct is cancelled, the lapsed subscribed can download files for a fee. However, even Elsevier cannot predict what may happen with content on Science Direct that comes from other publishers, although they are working on agreements (Hunter 2001). Kluwer gives four years of access after the conclusion of a subscription, and will provide an archival disk. Further, it allows subscribing institutions to make one copy of licensed material electronically for archival purposes. American Institute of Physics allows lapsed subscribers to purchase a physical archive copy of that term's material at the end of such a subscription term. The American Mathematical Society (http://www.ams.org/jourhtml/about.html) will provide a complementary disk containing all articles in the subscription year to subscribers of its electronic journals. Project MUSE provides a non-searchable file copy, usually on CD-ROM, containing all the articles published online during the previous subscription year, and MUSE allows local archiving even after a subscription is cancelled, because it presumes that libraries own the material from the electronic files to which they subscribe. The Royal Society of Chemistry allows continuing, perpetual access to the publications to which an institution had a subscription, even after cancellation.

Duranceau looked at how several e-journal providers addressed the issue of perpetual access and found that JSTOR allows ceased customers to receive a CD-ROM version of purchased materials after cancellation, that OCLC's Electronic Collections Online has all participating publishers sign an agreement allowing perpetual access to subscribed volumes for libraries even after cancellation, and that Blackwell's does offer continued access after cancellation, and is committed to the idea of perpetual access, but will not guarantee access because the publishers involved may decide otherwise, and Blackwell's thinks others should take on this service (Duranceau 1998). Duranceau also found that Highwire will not guarantee perpetual access, and does not offer a true archive, although they have free backfiles available, because it has not found a technology that captures the full capabilities of the electronic journals it offers, and because it has not found a simple cost-effective way to promise access to restricted online information (Duranceau 1998).

Thus, there seems to be no agreement on how to provide access to journals after a subscription is lapsed. Many publishers seem to think a static copy will suffice, but as the functionality of archives increases, this may be found to be too limiting. At that time some sort of extra access fee so that lapsed users can access the volumes for which they had subscriptions may become the norm.

CONCLUSIONS

The issues confronting electronic archiving of journals are numerous, but can be winnowed down to six main issues: (1) what is meant by electronic archiving, including what to archive and how long to archive, (2) who should do the archiving, (3) how should archiving be done, including all the technological issues, (4) how can integrity be protected, (5) how should access to archives be priced, and (6) what should policies be for access after cancellations or changes in publishers. The first question of what is electronic archiving has several parts that are mostly resolved, but there are still contentious issues of content and longevity that have no agreement at this moment. The question of who should do the archiving is extremely important because it reflects on all parts of the first question, and on the questions of pricing and access after cancellation. Third party solutions seem to be the favorite, but there are several projects in the works that may allow individual libraries to archive, or allow publishers to work with others to ensure archiving. The technological questions, although still vexing, seem to have found some consensus, and the viability of these solutions will depend on the establishment of, and compliance with, standards. The question of reliability seems to have some solutions, and some combinations of various digital watermarks and other technologies may be sufficient to assure authenticity. Pricing issues will become more important as more archives become available, but pricing will depend on who does the archiving, how it is done, what it includes and many of the other issues that remain unresolved. There seems to be agreement that there must be some sort of access to information after a subscription lapses, but there is no conformity in how that access is delivered at this time.

Thus, the issues that electronic archiving faces are mostly unresolved. However, there is considerable activity on the part of libraries, publishers and others to consider the issues and try to find solutions. Although one cannot be completely comfortable with the issues as they now stand, librarians who question the stability, reliability and future of electronic resources may obtain some reassurance from the extent of the efforts being made to find answers.

REFERENCES

Albanese, Andrew. 2001. Harvard, publishers to create E-archive. *Library Journal* 126 (11):16.

Anderson, Rick. 2000. Is the digital archive a new beast entirely? *Serials Review* 26 (3):50-52.

Anonymous. 1999. Preserving papers for posterity. *Nature* 397:198-199.

_____. 2001. Cornell project addresses the preservation of scholarly journals. Project Harvest. *Information Today* 18 (4):37.

Arms, William Y. 2000. Preservation of scientific serials: three current examples. Case studies of three electronic journals. *Microform & Imaging Review* 29 (2):50-6.

Boyce, Peter B. 1998. Costs, archiving and the publishing process in electronic STM journals. Experience of the American Astronomical Society. *Against the Grain* 10 (6):24-5.

_____. 2000. Who will keep the archives? Wrong question! *Serials Review* 26 (3):52-5.

Buckley, Chad, Marian Burright, Amy Prendergast, Richard Sapon-White, and Anneliese Taylor. 1999. Electronic publishing of scholarly journals: a bibliographic essay of current issues. (computer file). *Issues in Science and Technology Librarianship* 22. [Cited August 28, 2002. Available from http://www.istl.org/istl/99-spring/article4.html].

Butler, Declan. 1999. The writing is on the web for science journals in print. *Nature* 397:195-200.

Chepesiuk, Ronald. 2000. JSTOR and electronic archiving. *American Libraries* 31 (11):46-8.

Davidson, Lloyd A. 2000. The digital archiving of electronic journals: where we are and what problems lie ahead. A report of the LITA Electronic Publishing/Electronic Journals Interest Group meeting. American Library Association Midwinter Meeting, San Antonio, January 2000. *Technical Services Quarterly* 18 (2):60-9.

Day, Michael William. 1998. Online serials: preservation issues. *The Serials Librarian* 33 (3-4):199-221.

Douglas, Kimberly. 2000. Digital archiving in the context of cultural change. *Serials Review* 26 (3):55-9.

Duranceau, Ellen Finnie. 1998. Archiving and perpetual access for Web-based journals: a look at the issues and how five E-journal providers are addressing them. *Serials Review* 24 (2):110-15.

Feldman, Susan E. 1997. "It was here a minute ago!" Archiving the Net. *Searcher* 5:52-64.

Flecker, Dale P. 2001. Preserving scholarly E-journals. (computer file). *D-Lib Magazine* 7 (9). [Cited October 9, 2002. Available from http://www.dlib.org/dlib/september01/flecker/09flecker.html].

Friedlander, Amy. 2002. Digital preservation looks forward. *Information Outlook* 6 (9):12-18.

George, Sarah E. 2002. Before you cancel the paper, beware: all electronic journals in 2001 are not created equal. Experience at the University of Texas at Dallas. Report of a program at the 2001 NASIG Conference. *The Serials Librarian* 42 (3/4):267-73.

Gilson, Tom. 1997. Electronic and print: a merger in the making. Reference Reviews Europe Annual combines a print product with Web archiving and updating. *Against the Grain* 9:22+.

Goodman, David. 2001. One years' (sic) experience without print at Princeton (electronic journals). In *National Online 2001 (22nd:2001:New York, N.Y.). National Online 2001: Proceedings Information Today*. United States.

Graham, Rebecca A. 2000. Evolution of archiving in the digital age. *Serials Review* 26 (3):59-62.

Gyeszly, Suzanne. 2001. Electronic or paper journals? Budgetary, collection development, and user satisfaction questions. At Texas A&M University Policy Sciences and Economics Library. *Collection Building* 20 (1):5-10.

Harnad, Stevan. 2001. The self-archiving initiative. *Nature* 410:1024-1025.

Haynes, David, and David Streatfield. 1998. A national co-ordinating body for digital archiving? (computer file). *Ariadne (Online)* 15. [Cited December 20, 2002. Available from http://www.ariadne.ac.uk/issue15/digital/].

Hunter, Karen. 2001. Going "electronic-only": early experiences and issues. *Journal of Library Administration* 35 (3):51-65.

Hunter, Karen A. 2000. Digital archiving. At Elsevier Science. *Serials Review* 26 (3):62-4.

Incorporated, Inera. 2001. *E-Journal archive DTD feasibility study: prepared for the Harvard University Library Office for Information Systems E-Journal Archiving Project*. (computer file). [Cited August 21, 2002. Available from http://www.diglib.org/preserve/hadtdfs.pdf].

Kenney, Anne R., and Nancy Y. McGovern. 2001. *Cornell's Project Harvest: sub-ject-based digital archives.* (computer file). [Cited December 5, 2002. Available from http://www.library.cornell.edu/harvest/SBDA-CNI.ppt].

Keyhani, Andrea. 1998. Creating an electronic archive: who should do it and why? *The Serials Librarian* 34 (1/2):213-224.

Luijendijk, Wim C. 1996. Archiving electronic journals: the serial information pro-vider's perspective. *IFLA Journal* 22 (3):209-10.

Mabe, Michael A. 2001. Digital dilemmas: electronic challenges for the scientific jour-nal publisher. Presented at the DELOS workshop, November 1999, Pisa, Italy. *Aslib Proceedings* 53 (3):85-92.

Malinconico, Michael. 1996. Electronic documents and research libraries. *IFLA Jour-nal* 22:21-225.

Masanes, Julien. 2002. Towards continuous web archiving. (computer file). *D-Lib Maga-zine* 8 (12). Cited December 17, 2002. Available from http://www.dlib.org/dlib/december02/masanes/12masanes.html].

Montgomery, Carol Hansen, and Donald W. King. 2002. Comparing library and user related costs of print and electronic journal collections (computer file). At Drexel University. *D-Lib Magazine* 8 (10). [Cited December 17, 2002. Available from http://www.dlib.org/dlib/october02/montgomery/10montgomery.html].

Morris, Sally. 2000. Archiving electronic publications: what are the problems and who should solve them? *Serials Review* 26 (3):64-6.

Neavill, Gordon Barrick, and Mary Ann Sheblé. 1995. Archiving electronic journals. *Serials Review* 21:13-21.

Reich, Victoria Ann. 2002. Lots of copies keep stuff safe as a cooperative archiving solu-tion for e-journals. (computer file). *Issues in Science and Technology Librarianship* 36. [Cited December 5, 2002. Available from http://www.istl.org/02-fall/article1.html].

Reich, Victoria Ann, and David S. H. Rosenthal. 2001. LOCKSS: a permanent Web publishing and access system. Lots of Copies Keep Stuff Safe; Web cache for elec-tronic journals. (computer file). *D-Lib Magazine* 7 (6). [Cited October 9, 2002. Available from http://www.dlib.org/dlib/june01/reich/06reich.html].

Rogers, Michael. 1999. Librarians & publishers ponder preservation and archiving. at 1999 International Virtual Libraries Conference. *Library Journal* 124 (12):29.

Rowe, Richard R. 2002 Digital archives: how can we provide access to 'old' biomedical information. (computer file). *Nature Web Debates.* [Cited December 16, 2002. Avail-able from http://www.nature.com/nature/debates/e-access/Articles/rowe.html].

Solla, Leah. 2002. Building digital archives for scientific information. (computer file). *Issues in Science and Technology Librarianship* 36. [Cited December 5, 2002. Available from http://www.istl.org/istl/02-fall/article2.html].

Tagler, John. 1998. The electronic archive: the publisher's view. *The Serials Librarian* 34 (1/2):225-232.

Tenopir, Carol. 1999. Should we cancel print? Realities of digital periodicals. *Library Journal* 124 (14):138+.

Terry, Ana Arias. 2001. Digital archiving: a work in progress. *Against the Grain* 13 (3):28-30.

White, Martin S. 2001. Electronic access to scientific journals: problems, problems. *EContent* 24 (10):66-7.

APPENDIX A. Archiving Policies Survey

Does your company have a written policy on archiving electronic copies of print journals?

> a. If so, is it publicly accessible, and where can this policy be found?
> b. If not, are you planning to create such a policy and when do you anticipate its completion?

If you have an archival policy, or are in the process of creating one, does it address the following issues:

1. What type of group is (or should be) responsible for archiving?
   a. your company?
   b. your subscribers?
   c. another source (like JSTOR)?

If you answered b or c to question 1, please skip to question 10.

2. Who should pay for archiving?
   a. current subscribers since they are the ones who want the access to backfiles?
   b. your company since it is your company's product?
   c. others? If so, who–authors, copyright holders, professional societies?

3. If current subscribers should pay for archiving, how would that cost be structured?
   a. current subscribers would have access to backfiles included in their current annual subscription costs?
   b. current subscribers would have to pay extra access fees for backfiles, and could presumably decline access to backfiles to save money?
      –if so, would the extra fee be one-time or annual?

4. Should lapsed subscribers have some sort of access to the years for which they had subscriptions? If so,
   a. would this access be included in the cost of the subscriptions at the time they were active (that is, they paid for the data, therefore they get access, versus they rented the data and only got access while paying for it)?
   b. would this access require some additional fee?
      –if so, would it be a one-time fee or an annual fee?
   c. have you considered CD-ROM access so that lapsed users could have access to the backfiles but they would then be responsible for maintaining that access?

5. How much access would be granted? What would be terms of a subscription?
   a. should current subscribers have access to a limited rolling file?
   b. should current subscribers have access to all years for which they had subscriptions?
   c. should current subscribers have access to all the backfiles, or only years for which they had subscriptions?

6. How will you prevent the data from becoming obsolete? When will you migrate data?
   a. will you migrate data as soon as new technologies are available?
   b. will you wait until the technologies in use are almost obsolete?
   c. will you migrate on a regular schedule regardless of the technology available?

7. How will you ensure uncorrupted data?
   a. will the print version continue to be the version of record?
   b. what mechanisms have you considered to prevent corruption/alteration of the data–do you assume PDF files will be incorruptible?
   c. will the archives reside on more than one computer, and in more than one geographic region?

8. Have you considered what would happen to your archives if:
   a. your company was bought by another company?
   b. your company ceased to exist?

9. How many staff members would be trained and have access to manipulate and update the archives?
   a. 1-5?
   b. 5-10?
   c. 10+

10. Have you made any agreements with other services (like JSTOR) to archive your data?
   a. if so, have you looked into standardized DTDs (document type definitions)
   b. would you partner with an academic institution, like Harvard's E-Journal Archiving Project?

User Expectations
and the Complex Reality
of Online Research Efforts

David Stern

SUMMARY. Serious research involves a complex variety of efforts. Critical thinking is supported and supplemented through the skilled searching, tracking, retrieving, and manipulation of previously created material. The research process has been significantly streamlined and enhanced by the development of online tools, but comprehensive research still requires manual labor. This article will compare current user expectations to the realities of present online research capabilities. It will discuss new supporting technologies, some of the most often observed problems in using the present online research tools, user misconceptions about these tools, and the efforts libraries must make to raise the awareness of the possibilities and limitations to online-only research. The example of a changing (enhanced) linking technology option will demonstrate the need for researchers to adjust their expectations–to understand and accept the differences between previous "ease of access" library services and new "complex research navigation and exploration" options. *[Article copies available for a fee from The Haworth Document Delivery Service: 1-800-HAWORTH. E-mail address: <docdelivery@haworthpress.com> Website: <http://www.HaworthPress. com> © 2002 by The Haworth Press, Inc. All rights reserved.]*

KEYWORDS. Research process/methods, online research, enduser expectations and behaviors, technology enhancements, linking technologies, critical thinking, limitations of technology

David Stern is Director of Science Libraries and Information Services, Kline Science Library, 219 Prospect Street, P.O. Box 208111, New Haven, CT 06520-8111 (E-mail: david.e.stern@yale.edu).

[Haworth co-indexing entry note]: "User Expectations and the Complex Reality of Online Research Efforts." Stern, David. Co-published simultaneously in *Science & Technology Libraries* (The Haworth Information Press, an imprint of The Haworth Press, Inc.) Vol. 22, No. 3/4, 2002, pp. 137-148; and: *Scholarly Communication in Science and Engineering Research in Higher Education* (ed: Wei Wei) The Haworth Information Press, an imprint of The Haworth Press, Inc., 2002, pp. 137-148. Single or multiple copies of this article are available for a fee from The Haworth Document Delivery Service [1-800-HAWORTH, 9:00 a.m. - 5:00 p.m. (EST). E-mail address: docdelivery@haworthpress.com].

Every library must delicately and selectively raise and fulfill library user expectations based upon their mission and stated research intentions. Due to both technological and budgetary barriers, most libraries cannot hope to satisfy all user expectations. On one extreme of user expectation are the interests of the casual searcher who only requires easy access to facts or full-text materials. On the other extreme is the serious researcher requiring complex navigation and exploration options in order to comprehensively cover the full range of existing materials. This article will discuss user expectations, online research support technologies, the resulting frustrations when services do not match expectations, the often unknown inadequacies of research methods when only online tools are used, and the efforts libraries must make to raise the awareness of the possibilities and limitations of online-only research. The ultimate goal is to have all users understand the place of these online tools in the complete research process, to have users provide feedback for revisions, and have researchers eventually accept the power, complexity, and limitations of existing services.

NOVICE RESEARCHER EXPECTATIONS

Many new researchers expect the computer to perform all required information discovery tasks. These researchers may not differentiate between the convenience of casual information surfing and the more complex requirements of in-depth research activity. Novice researchers often believe that the tedious efforts of the past can be accomplished by the fast processing of a networked computer. Or if they don't actually believe, they are willing to extend belief for the sake of speed and convenience.

Serious research has always involved effort, creativity, and critical thinking. In the past, readers expected to spend hours mining the previous literature in order to discover seminal arguments and supporting data. These foundations were the fodder for new insights. The serendipitous aspects of research also produced unexpected and unanticipated results. While research was laborious, it was exciting and productive.

With the advent of indexing and linking technologies, the effort involved in efficiently identifying and obtaining previous work was reduced. Indexing and abstracting tools allowed researchers to quickly filter through the growing body of literature using key concepts, then keywords, and finally combinations of key terms with limits such as years, languages, and full-text availability. The linking to and delivery of full-text to reader desktops allowed for far more convenient and rapid research.

However, efficiency is not the same as effectiveness. Rapid and convenient online access is not equivalent to comprehensive research. Not all items are identified using online search tools, and many items are still not available in full-text. Observation has shown that many researchers will simply ignore important literature if it requires a walk to a physical library. For many readers convenience has become more important than quality or comprehensivity.

This preference for convenience is partially the result of librarian actions. Librarians collaborated on the creation, and then strongly advertised, a system, which emphasized easy access. For many readers this is a perfectly appropriate approach to research; a casual reader often only wants a few articles about many topics.

For a serious researcher, on the other hand, these technologies often do not adequately substitute for complete research processes. These online technologies only partially address the discovery process, and comprehensive discovery still requires old-fashioned efforts. In particular, there is no online replacement for critical thinking and creativity. Researchers still need to explore a wider variety of materials as part of their critical thinking process. Serious research certainly involves reviewing previously published thoughts and tracing the development of ideas over time using imbedded references. However, it also involves scanning the literature in order to find important but not already/specifically identified related ideas. This environmental scanning of the literature is done for both general educational growth within a discipline and as episodic subject-specific exploration. One cannot expect an index to identify all relevant material; each index has its limits in terms of depth, scope, and size. Serious research still requires significant manual effort for the discovery of all related materials.

Whether it is always appropriate or not, searching non-peer reviewed material is becoming an ever more popular and important way to discover many types of information (Weston, 2002). Searching the Internet provides many important and useful facts and perspectives, but researchers must understand the underlying validity and sources of this information. All too often data retrieved from the Internet are accepted as absolute and current without any critical analysis. This behavior must be challenged using examples of false data and comparisons with peer-reviewed alternatives.

USER BEHAVIORS IN RELATION TO TECHNOLOGIES

Library user behaviors have always demonstrated that easier discovery and access methods are desirable and utilized when offered. Users adopt tools that make research easier almost immediately; think of online book catalogs, online journal table-of-contents delivery, and links to full-text. Even complex enhanced options are adopted quickly when relevant; think of CD-ROM self-searching, saved search strategies, and current awareness services. The delay in adoption is often directly related to the actual recognition of the value of the enhancements. Experience has shown that the recognition of value often comes through demonstrations, not discussions of possibilities. This is seen in visits to faculty offices, in which demonstrations of cited searches to their own works make instant converts out of people who never used these tools (even after numerous announcements and conversations). Demonstrations in classroom and laboratory settings are also quite successful in creating users. Just as in storytelling and writing, showing is far more impressive than telling.

Rather than simply being easier than paper-based tools, online research tools are becoming ever more powerful—and therefore more complex. In reaction to the natural reticence to learn new tools and techniques, librarians need to develop user comfort with, and acceptance of, this complexity (Brandt, 2001; Ren, 2000). This acceptance will occur as a direct result of raising the awareness of the value of these advances and enhancements for individual user needs. Users must also recognize the limitations of these tools and accept the additional manual effort that is required for good research results.

In short, we will need to convince serious scholars (1) that it is time to value exploration possibilities over convenience services, (2) to learn the more complex online search methods in order to maximize the possibilities of online research tools, and (3) that there are limitations to online research that still necessitate manual research efforts.

FACTORS IN CHANGING USER ATTITUDES AND BEHAVIORS

Discovery

The first portion of the research process that most users explored with computers was that of search and discovery. Computers have always been seen as excellent tools for searching text; think of keyword searching in databases (i.e., dictionaries, thesauri, and documents) and find-and-replace tools in word processors. The excellent MARS bibliography, *"Users' information-seeking behavior: what are they really doing? A bibliography"* (2001), documents many typical user behaviors and provides excellent readings for many aspects of this complex process. Among the major misconceptions has always been the assumed data set against which one is searching: everything at one time (Grimes 2001; Kibirige, 2000; Bates, 1996). Often user expectations of computer discovery far exceed that of actual systems–many researchers still believe that web search engines such as Google and Yahoo will find all items on the Internet, or that searching library online book catalogs will find journal articles. Perhaps federated search engines will address these possibilities in the future, but for now the reality is that most searches are run against rather limited domains. This domain limitation must be stressed to novice users.

Even if one clearly understands the exact scope and time coverage of a resource, there are many other research issues that must be considered when searching indexes, full-text, or other information sets. One major issue is search strategy development, which is often not well managed when performing searches. Search statements often include significant structural inadequacies; and garbage in-garbage out is still the rule with computers. The most obvious search behavior that needs improvement is that of addressing language complexity. A large percentage of search strategy problems involve the lack of truncation and synonyms, and the use of (perhaps unintended) exact phrase searching instead of far less restrictive Boolean combinations of text. Users must be taught to use appropriate search strings and proximity operators, challenging their expectation that the computer can do all the manipulations, analysis and cross-referencing.

Users must also be taught the special requirements of simultaneous searching within heterogeneous databases. When searching multi-subject databases, or more than one database at a time, researchers must consider the use of multiple subject headings, and/or classification scheme normalization. There is also the issue of appropriate duplicate record removal; records from certain databases always contain additional data and should be the preferred result.

Using natural language search software also has its own built-in expectations and hidden processes that must be understood if researchers expect to perform serious searching. Knowing what is happening behind the black box is important if one is to perform true interactive research. One cannot critically analyze results and revise searches if unaware of the underlying search statement criteria. Therefore, even natural language searching must be understood at some basic level rather than simply trusted unconditionally.

Ever more advanced search options are appearing, and these require even more sophisticated search input (Stern, 1999). Powerful image analysis search engines are now available for finding similar shapes and colors. Visualization tools allow for novel discovery methods by processing and displaying search results in alternative ways that may help in identifying best areas for analysis. In many cases users may not be aware of what is happening during their searches, but they should be able to view/understand some degree of their search strategies if they are to document their research approaches.

Manipulation

Another significant change in the research process has been the ability to manipulate data, both raw data and descriptive metadata. This allows for new discovery and real-time analysis options as well as additional post-search data storage, retrieval and display possibilities. User behaviors and expectations are changing radically due to this ability to manipulate data.

The increasing use of metadata has allowed for powerful new research methodologies. Many search services offer these new techniques, for instance: "find similar" options; citation tracking links; frequency generated pathways (e.g., Google links); related records identification; searches of images by keyword, shape, and color; current awareness notification services; expert systems assistance; and visualization displays. The development of advanced metadata searching algorithms is underway for improved federated searching of web-based materials (Lawrence, 1999).

GIS software is one example of a powerful combination of these possibilities. Various types of data are combined and result in new and impressive research results, in both graphic and data set formats. In these instances, the discovery of geospatial information is frequently the most difficult part of the research process. Often local level data must be created rather than discovered. Federated resource clearinghouses for this commercial and locally created data would make GIS data sharing much more efficient. The gap between library coordination of existing data and user expectations of complete data repositories is significant and must be stressed to researchers. Even with serious attention, a great deal of this locally produced data (often by amateurs) will never be consolidated into central repositories. The appropriate statistical use of this raw GIS data is another important critical thinking issue that is beyond the scope of this article.

The increasing use of personal database software such as EndNote allows for the creation of personalized researcher knowledge databases. These knowledge management packages offer powerful new storage and search options for individual researchers. Advanced local databases such as ProCite and DB Textworks offer even greater possibilities for integrated media. Databases such as www.scopeware.com offer novel (in this case time-based) approaches for organizing personal data. Sophisticated database creation is often required, with complexity being the cost to users for these new and powerful tools. Researchers are developing specialized behaviors and new expectations through the use of these tools.

Libraries offer new services based upon this same data manipulation; examples include journal citation delivery services and book catalogs, which include links to online book reviews. These new possibilities should be emphasized in orientation and instruction sessions (McFadden, 2002; OCLC White Paper, 2002).

Linking

The advent of linking between documents has significantly changed user expectations and behaviors. At first users expected only convenient links to full-text documents, but these expectations are beginning to expand as services are now offering additional linking options. These newer linking options provide access to related services and resources such as automatic author searches (in journals and book catalogs), citation tracking, and Internet item identification (i.e., equipment, supplies, reviews, and upcoming events). Additional linking options on the horizon will include features such as multi-directional links and seamless multimedia integration.

Users will probably expect these enhanced links to cover all possible items in much the same way that they currently expect Internet searches to cover all materials. The reality is that these links will initially also cover a limited domain of resources. Federated searches and resulting links are not always comprehensive, and certainly not in the early stages of metadata linking across the Internet. Adoption of standard protocols and tools (such as OAI servers) will move us farther along the path of automatically identifying and linking non-standard classification and metadata repositories (i.e., images, audio, etc.), but it will be quite some time before these automatic links will cover the broad set of resources that users may already expect.

EXAMPLE OF ENHANCED LINKING

One example of a new tool in which search results differ from user expectations is the SFX linking found between information Resources (search tools) and Targets (documents). While users of library indexes have come to expect on-screen links to full-text materials, SFX presents an alternative on-screen option which is not what many people are accustomed to seeing as a result. The SFX service does not present immediate full-text availability information, but instead offers an on-screen button which produces item-specific enhanced linking options when activated (Figure 1). For many casual researchers the convenience of immediate full-text link displays is adequate, and all that they desire. These immediate on-screen full-text links, as offered by aggregators and indexes with pre-populated subscription information, can only appear if you duplicate your subscription database in every search platform, and this is no longer reasonable for libraries that utilize a number of search services.

This difference in the immediate display elements is the result of scalability issues, given all the new linking possibilities. It is far more efficient to only check for so many possible links to an item on a need-to-know basis. In order to efficiently handle so many related item possibilities, it would be unreasonable to pre-check every citation against all potential links (i.e., a central subscription database, author citation searches, Internet search engines, and more). In such a large and complex networked environment, in terms of user response time, it is best if potential related links are only ascertained when users ask for full information on specific references. This new enhanced service approach requires acceptance of different types of effort and intermediate results in order to provide enhanced service.

For serious researchers, expanded options and possibilities must become a higher priority than the immediate identification of full-text materials. Many researchers will certainly want links to as much related information as possible . . . but they have not yet seen examples in order to understand these options and to accept the new and different efforts. It is our task to demonstrate that the long-term benefits to such changes are well worth the frustration of yet another learning curve.

OTHER VENUES OF CHANGE

Researchers are also seeing changes to their information management behaviors in other related scenarios. The online distribution of

FIGURE 1. SFX Enhanced Link Options from a Journal Index Database

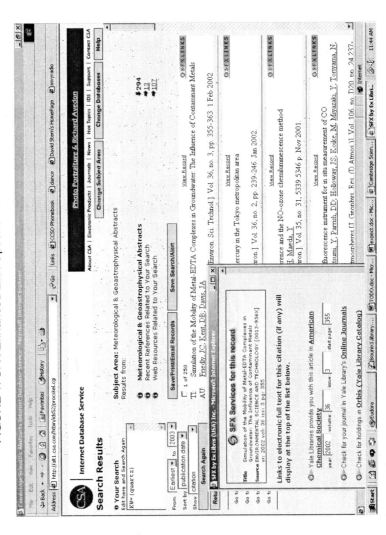

Used with permission.

both peer reviewed and other materials is occurring in many new ways that may require changes in behaviors, technologies, and expectations. In some cases these predicted changes are not occurring as quickly as once predicted (Maughan, 1999). These new tools and techniques also influence teaching pedagogy.

Publishing

Significant changes are occurring in the world of publishing due to changes in the underlying online information environment. In traditional journal publications there are changes in the submission and peer review processes. In this new world many authors are required to provide far more sophisticated digital material to publishers. Peer review occurs much more quickly, with a much shorter time lag between submission and publication. At the same time, in selected fields this delay is still not acceptable and alternative e-print servers exist to distribute pre-peer-reviewed materials. Although it may be transparent to many users, some journal materials are now available at no cost to libraries. Unfortunately, many readers wrongly imagine that all online journals are free on the Internet; this is a belief libraries must correct using both branding and education about journal costs. It would also be in the best interests of libraries, authors, editors, and readers to free the distribution of online journals from the expensive subscriptions of commercial publishers, but this is a related topic requiring extended conversations elsewhere.

Many online journals now include enhanced features such as links to supplementary integrated media, links to outside journals and databases, citation tracking, current awareness notification, table-of-contents delivery, and personal virtual file cabinets. Some of these services require users to identify themselves to the journal using passwords and/or e-mail addresses. Privacy may become an issue in certain cases, and many users are unwilling to provide this personal data but still expect such added-value services. Personalized services (i.e., MyLibrary profiles, customized portals, individualized news feeds) must be maintained without endangering privacy.

Online book services are not yet widespread due to impractical pricing models, technical problems, and user dissatisfaction with reading online; but the early systems do provide enhanced features such as keyword searching, cut-and-paste options, highlighting, commenting, and citation links. Surely, there are advantages to such systems for certain applications, and researchers will want these options once the online book becomes a reality. It is only a matter of time before researchers expect these enhanced options for online books. Librarians must work with ebook vendors and services to develop these powerful tools within reasonable pricing models and scalable annual selection processes.

Classrooms

Significant changes are also occurring in the worlds of classroom and laboratory teaching. Individualized, self-paced, and non-linear pedagogy is now possible using online tutorials and web-based presentation software. Integrated media (i.e., imbedded sound and video bytes, programming, and computer simulations) are now possible in many classrooms. Automated online testing and grading is also possible. New and enhanced learning possibilities exist, but this requires new teaching tools, the creation of new materials, and changes in the behaviors of both instructors and students. Librarians should be proactive collaborators with teachers (and students) in order to develop standardized teaching tools that can be archived, searched, retrieved, and reused well beyond the life of the first specific teaching intention, within current copyright guidelines.

CONCLUSION

It is clear that there are significant changes in the research process due to online tools. It is also clear that these tools are still changing and this requires ongoing user training. These online tools provide great advantages in terms of time and effort saved, but they cannot completely replace manual research. For serious researchers, expectations need to be modified to (1) accept that extra critical thinking efforts are necessary in order to maximize the explorations required for quality online research, and (2) recognize that researchers will always need to supplement online research with manual research in order to accomplish relatively comprehensive investigations.

REFERENCES

Marcia J. Bates, "Learning about the Information Seeking of Interdisciplinary Scholars and Students" *Library Trends* 45: 155-164 (Fall 1996).

Brandt, D.S. Information technology literacy: task knowledge and mental models. *Library Trends* 50 (1): 73-86 (Summer 2001).

Grimes, D.J. et al., Worries with the Web: a look at student use of Web resources. *College & Research Libraries* 62 (1): 11-23 (January 2001).

Herring, M. "10 reasons why the Internet is no substitute for a library" *American Libraries* 32: 76-78 (2001).

Kibirige, H.M. et al., The Internet as a source of academic research information: findings of two pilot studies [of 4 New York academic institutions]. *Information Technology and Libraries* 19 (1): 11-16 (March 2000).

Lawrence, S. & Giles, C. "Searching the Web: General and scientific information access" *IEEE Communications Magazine* 37: 116-123 (1999).

Maughan, P.D. Library resources and services: a cross-disciplinary survey of faculty and graduate student use and satisfaction [at the University of California at Berkeley]. *The Journal of Academic Librarianship* 25 (5): 354-66 (September 1999).

McFadden, T.G. Computer-based instruction in libraries and library education [special issue]. *Library Trends* 50 (1): 1-158 (Summer 2001).

OCLC White Paper on the Information Habits of College Students "How academic librarians can influence students' web-based information choices" (June 2002) (http://www2.oclc.org/oclc/pdf/printondemand/informationhabits.pdf).

Ren, W. Library instruction and college student self-efficacy in electronic information searching [survey of 85 undergraduate students taking English composition at the Newark campus of Rutgers University]. *The Journal of Academic Librarianship* 26 (5): 323-8 (September 2000).

Stern, David. "New Search and Navigation Techniques in the Digital Library" *Science & Technology Libraries* 17 (3/4): 61-80 (1999).

Users' information-seeking behavior: what are they really doing? A bibliography. *Reference & User Services Quarterly* 40 (3): 240-50 (Spring 2001).

Weston, Wil. "Access to scientific literature" *Nature* 420 (19):19 (November 7, 2002).

Bibliometric and Bibliographic Analysis in an Era of Electronic Scholarly Communication

Locke J. Morrisey

SUMMARY. Bibliometric analysis of citation data is important to scientist and librarian alike. With alternate means of scientific scholarly communication proliferating, it's important to be able to accurately link publications and their references. This article highlights some of the current problems that arise when doing citation analysis of different kinds of scientific scholarly communication. A combination of better bibliographic

Locke J. Morrisey, MLS, is Head of Collections, Reference & Research Services, Gleeson Library/Geschke Center, University of San Francisco, 2130 Fulton Street, San Francisco, CA 94117-1049 (E-mail: morrisey@usfca.edu).

[Haworth co-indexing entry note]: "Bibliometric and Bibliographic Analysis in an Era of Electronic Scholarly Communication." Morrisey, Locke J. Co-published simultaneously in *Science & Technology Libraries* (The Haworth Information Press, an imprint of The Haworth Press, Inc.) Vol. 22, No. 3/4, 2002, pp. 149-160; and: *Scholarly Communication in Science and Engineering Research in Higher Education* (ed: Wei Wei) The Haworth Information Press, an imprint of The Haworth Press, Inc., 2002, pp. 149-160. Single or multiple copies of this article are available for a fee from The Haworth Document Delivery Service [1-800-HAWORTH, 9:00 a.m. - 5:00 p.m. (EST). E-mail address: docdelivery@haworthpress.com].

control, interactive systems, and adherence to standardized electronic publishing protocols would improve the accuracy and reliability of the citation data retrieved. *[Article copies available for a fee from The Haworth Document Delivery Service: 1-800-HAWORTH. E-mail address: <docdelivery@haworth press.com> Website: <http://www.HaworthPress.com> © 2002 by The Haworth Press, Inc. All rights reserved.]*

KEYWORDS. Bibliometrics, citation analysis, bibliographic control, scholarly communication

Research scientists and engineers have been fascinated by citation analysis, or bibliometrics, ever since Eugene Garfield's first article came out in the mid-1950s, proposing a "bibliographic system for science literature that can eliminate the uncritical citation of fraudulent, incomplete, or obsolete data by making it possible for the conscientious scholar to be aware of criticisms of earlier papers" (Garfield 1955). Only thereafter, people began to evaluate research by looking at citation data (Martino 1967; Is Your Lab Well Cited? 1970). Academic science and engineering librarians have long seen the rush to perform a *Science Citation Index* (now *Web of Science*) search when faculty are preparing promotion and tenure packets, thinking that numerous references to their publications may somehow tip the scale in their favor. Librarians have also used some of these data to make informed choices when it comes to reanalyzing journal subscriptions (Garfield 1994; Sylvia 1998; Lascar 2001). But with the explosion of alternative methods of scholarly publishing that are freely searchable on the Web (Odlyzko 2001), there need to be changes in the way one evaluates and searches for bibliometric data.

WHAT IS BEING COUNTED, WHAT COUNTS AND WHAT DOESN'T?

There are a few controversies that swirl around the use and accuracy of citation data. Traditional citation indexes only compiled data from the journals they chose to index; therefore, much of the citation data that were included in conference proceedings papers, technical reports, working papers, book chapters, etc., were never counted. Even the accuracy of the standard commercial product, *Science Citation Index*, has been brought into question (Errors in Citation Statistics 2002). Many believe it's important to look at who's doing the citing, so numbers of citations are not artificially inflated with authors citing their own work or having co-authors cite their work. With the advent of preprint servers, online tech report archives, and electronic autonomous citation indexing (ACI) systems such as NEC Research Institute's *CiteSeer* (Giles 1998), an enormous amount of additional citation data are now available.

If one considers publications such as online bibliographies, reading lists, and course syllabi, a whole new set of potential citation data comes into play that was heretofore uncountable. So how does one collect and then analyze the citation data from all of these electronic sources? Some may believe that a faculty member's work listed as required reading on a university instructor's syllabus has the same amount of significance, although different, as a reference listed along with 100 others from a peer-reviewed journal article. On the other hand, some look at a citation from a peer-reviewed journal as being the gold standard vis-à-vis having more weight than references from other sources (Members of the Clever Project 1999).

OUT OF [BIBLIOGRAPHIC] CONTROL

With the increase in freely available, full-text research papers appearing online, outside the realm of the traditional indexing sources or aggregator databases, there is less authoritative quality control and a higher likelihood of differences or errors in the way articles are being cited. Looking at the ways one of this author's past publications has been cited on the freely accessible Web is a good example (see Figure 1).

By looking at the authors' names, one can see that each citation is slightly different. Even though autonomous citation indexing tools are good at pulling together citations that may be presented differently (Lawrence 1999), they are not foolproof. Imagine someone putting in all of the possible author combinations above to pull up publications where the reference is cited. With the addition of citation data from freely available online publications come more opportunities for errors. This is particularly true if the large numbers of student research papers that end up on the Web are included in citation analysis. In the five citations in Figure 1, the second one cites the wrong volume and end page for the article, and the last one gives credit to three authors, "Locke, Morrisey and Case," within the text of the publication instead of just two. This also brings into question how to weigh citations that are com-

FIGURE 1. Differerent Ways the Same Article Has Been Cited

1. Morrisey, LJ and DO Case. "There Goes My Image: The Perception of Male Librarians by Colleague, Student, and Self." College and Research Libraries, v49, no. 5, pp. 453-64 (Sep. 1988).
2. L.J. Morrisey & D.O. Case (1988) "There goes my Image: The Perception of Male Librarians by Colleague, Student and Self." College and Research Libraries, 48, 5:453-463.
3. Morrisey, Locke J. and Donald Owen Case. "There Goes My Image: The Perception of Male Librarians By Colleague, Student, and Self." College and Research Libraries 49 (September 1988): 453-464.
4. Morrisey, Locke J., and Donald O. Case. "There Goes My Image: The Perception of Male Librarians By Colleague, Student, and Self." College and Research Libraries 49 (September 1988): 453-64.
5. Locke, Morrisey and Case . . . (within the text of the article).

ing from student papers. Should one count citations from white papers or senior theses but not from upper division term papers? Who's to say?

TRACKING DOWN A MISCITATION

Besides coping with the nuances of the ways authors are referenced, science and engineering librarians are used to tracking down citations where the date, pages, or journal name listed in a reference are inaccurate. Consistent incidences of miscitations have been documented in studies that cover several disciplines (Eichorn 1987; Sweetland 1989; Rudolph 1990; Stull 1991). In the past, performing a quick database search or referring back to cumulative indices usually identified the correct citation. With cross-linking citation products such as *CrossRef*, it should be much easier to verify the accuracy of e-journal article references. In addition, *CrossRef* just announced, at the Annual Association of American Publishers Professional and Scholarly Publishing Conference held in February 2003, that they are beginning to incorporate e-books and other electronic non-serial formats into their protocols (Needham 2003). But if the publication is not from a standard e-journal or does not follow *Open Archives Initiative (OAI)* protocols, verification could get tricky. The URL for an electronic publication can be long and cumbersome. Mistyping the URL is very unforgiving. If only one letter, number, period, or slash is off, one will get the dreaded "404 Error." It's much easier to verify an incorrect URL if it is laid out in a systematic way, e.g., indicating an accession or report number in the URL extension. If there is no systematic arrangement, tracking down the current URL could be difficult, especially if the title of the paper is not included in the citation, a practice not uncommon in scientific publishing. One needs to determine if the URL was once correct and now has a different URL, or if the publication is no longer available online.

MULTIPLE URLs

Another digital dilemma in analyzing citation references is how to handle multiple URLs for the same reference. How are they counted? Are they cross-listed? Is there a definitive URL? Most commercial publishers are identifying their electronic content through Digital Object Identifiers (DOIs), but this is not usually the case for electronic publications that reside on educational or governmental web sites (see Figure 2).

As complex as the scenario appears, both McMaster and the CCHE have done an admirable job in trying to refer the user to the appropriate content. Where things really start to break down is in trying to answer some of the questions in Figure 3.

"HOLLOW CITATIONS"

At the Special Libraries Association (SLA) 2002 Conference in Los Angeles, the author attended the 7th Quadrennial Trisociety Symposium co-sponsored by the Chemical Division of SLA. One presentation was

FIGURE 2. Multiple URLs for the Same Web Page

1. Canadian Centre for Health Evidence (CCHE) "Users' Guides to Evidence Based Practice." The current URL is http://www.cche.net/usersguides/main.asp
2. URL from a reference that this author had cited in December 2001: www.cche.net/principles/content_all.asp This appears to be the exact same site, without an intermediary page listing a referring URL. But the current page now has a different title than the December 2001 citation.
3. Using a hotlink off a McMaster University web page: http://hiru.mcmaster.ca/ebm/userguid/userguid.htm URL refers to a page with the message "Online materials to support teaching of the Users Guides are not available until further notice. EBM Informatics Editors" with no referring URL.
4. Truncating that URL at: http://hiru.mcmaster.ca/ebm/userguid This gives the message "Online materials to support teaching of Evidence-based Health Care, including the Users Guides to Evidence-based Practice are now supported through the Centres for Health Evidence at: http://www.cche.net, in the 'Principles' section. The older version of the guides is maintained by the Centres at: http://www.cche.net/ebm/userguid --EBM Informatics Editors"

FIGURE 3. Questions About Bibliometrics and Multiple URLs

Did past versions of the URL point to different content?
Is the content the same, but just relocated on the host server?
How many different URLs being referenced exist, especially mirror sites, and how are they accurately counted *in toto*?
How does one measure the number of times an article is cited if the print version is cited differently than the electronic?

the report of a study regarding e-journal versus print journal usage in a chemistry library collection. When asked if a bibliometric analysis of faculty's publications had been done to see if the journals faculty were citing had any correlation to in-house journal use statistics, the presenter acknowledged there was interest in doing a citation analysis but not the opportunity to do so. What this question alludes to is that some faculty will write research papers citing articles that they have never read or even seen. This is what I refer to as a "hollow citation" or "hollow reference," having no physical substance, i.e., no copy of the article, behind it. How does one know this practice occurs? One way is that when an inaccurate reference shows up in a citation index with more than one article citing it, the odds are that the second and following citing authors took the inaccurate reference from the first citing article's bibliography and not from the article itself. Not only does this corrupt the citation database, but also it brings up issues of what some see as potential misconduct, if not plain laziness, in the scientific community. The U.S. Office of Research Integrity (ORI) has recently raised this concern in a proposed survey of NIH grant applicants (Soft Responses to Misconduct 2002; Ball 2002). The survey has been greeted with grumbling from a couple of scientific societies as being too nitpicky. How frequent is this practice? The answer may lie in a recent article written by Simkin and Roychowdhury (2002). The electrical engineering professors from UCLA have proposed a method based on stochastic modeling that estimates the percentage of authors who've actually read an article before they've cited it. It should be noted that the authors "... adopt a much more generous view of a 'reader' of a cited paper, as someone who at the very least consulted a trusted source (e.g., the original paper or heavily-used and authenticated databases) in putting together a citation list." The results are rather alarming. Using their model, they estimate that "only about 20% of citers read the original [article]" and "conclude that misprints in scientific citations should not be discarded as mere happenstance, but, similar to Freudian slips, analyzed."

ONLINE ONLY?

Publisher-based electronic journal packages have been available to libraries for several years now. These so called "Big Deals" (Frazier 2001) usually involve some type of library consortia and are renegotiated every 3-5 years. With a downturn in the economy and collections budgets being tight, librarians are beginning to think about discontinuing or at least scaling back some of these expensive electronic journal packages offered by for-profit publishers. At the 2003 American Library Association (ALA) Mid-winter Meeting in Philadelphia, the Scholarly Publishing and Academic Resources Coalition (SPARC) and the Association of College and Research Libraries (ACRL) joint forum, Theodore Bergstrom, chair of the economics department at the University of California at Santa Barbara, presented statistics that showed that "librarians were now spending 91 percent of their serials budgets on for-profit journals, which in turn account for just 38 percent of citations" and gave a short economics lesson to librarians as to "why site licenses for bundles offered by for-profit publishers, while offering wide access, actually harm the buying power of libraries" (SPARC/ACRL Session 2003). Collection development librarians can now take Bergstrom's words of wisdom combined with the usage data from these packages to see what their faculty and students are using and how cost-effective these package plans really are. With literally hundreds of thousands, if not millions, of dollars being spent for online access to thousands of journals, will the scientist or engineer bother using literature that is not online and easily accessible? If one of these packages is dropped or the content is changed significantly, will the author default to mostly referencing the materials that are electronically on hand or possibly rely more heavily on "hollow citations"? Studies by both Broch (2001) and Lawrence (2001) demonstrate that with increased accessibility to scientific literature online, authors are much more likely to cite an article that is freely accessible versus one that is not. Will an increase in the usage and citation of only freely or easily accessible articles somehow result in a "dumbing down" of the research process as well as skewing of citation data?

TAKING BACK BIBLIOGRAPHIC CONTROL

Information systems exist that can improve bibliographic accuracy as well as track bibliometric data better than the ones currently in place. The problem is that not all systems talk to each other yet. One of the more obvious needs is for comprehensive standardization across different electronic publishing platforms so articles can be cross-referenced and tracked more effectively. As more and more e-publishers become OAI compliant, this should get better over time. But there are still a couple of major hurdles that the scholarly scientific publishing community needs to overcome if major headway is to be made in this area.

One is how to get around the variety of ways author names appear in references. If the e-publishing community could create a public domain Uniform Author Identifier (UAI) similar to a DOI, then this UAI could be used within and across a multitude of online databases or e-journal collections. The variations on this author's name in the first example were minor compared to others in the literature: use of middle initial or not, changing surnames, dealing with common surnames and sets of initials, hyphenation or non-hyphenation, variation in transliterations from non-Roman to Roman alphabet characters and even the inability to determine the given name from the surname of the author. This last point may seem odd until one realizes that authors from Asia often list their names as "surname given name" rather than the Western tradition of "given name surname." So if one searches on an assumed last name and first initial, one may get a strange result. Worse still, sometimes this information changes depending on where the article is published. The same Asian author may have listed "given name surname" in the reference if it's known that the article will be appear in a U.S. or UK publication. As it is now, one may have to search both ways to accomplish a comprehensive search. A UAI would help avoid such confusion. It would establish another access point that could be used to link across databases using *SFX* or similar technology (Walker 2002) to more accurately retrieve author search results. The UAI could be numeric or alphanumeric to avoid some of the confusion alluded to above.

Going one step further, is there a way to create a Uniform Concept Identifier (UCI) that would be searchable across a number of databases? Some would argue that for most researchers, machine-generated keyword searching is all that is needed for looking for electronic publications by topic. There is a good argument to the contrary (see Figure 4). Note the differences in the thesaurus terms used for the concept of "posture," e.g., slouching versus good posture, in four database thesauri.

Not only do the databases not define the concept the same, but also the terminology is used differently depending on the database. The National Library of Medicine (NLM) has been working on developing a *Unified Medical Language System* (*UMLS*), a type of metathesaurus,

FIGURE 4. Comparison of the Concept "Posture" Across Four Database Thesauri

All thesaurus terms are **bolded**
MEDLINE: **Posture** (includes standing, sitting, patient positioning or how the body is carried) narrower terms include **Supine Position** and **Prone Position**
CINAHL: **Patient Positioning** narrower terms include **Supine Position** and **Prone Position** related terms include **Standing** or **Sitting** The term **Posture** is used separately to refer to how the body is carried.
PsycInfo: **Posture** (used for both standing, sitting, patient positioning or how the body is carried) related terms include **Body Language** and **Physique**
Physical Education Index: **Posture** (used for how the body is carried) Separate but unrelated terms are **Standing** or **Sitting**

for some time now (U.S. National Library of Medicine 2003). One of the problems they've run up against is that a number of existing thesauri are proprietary and would have to be licensed if they were to be used (e.g., thesauri from *CINAHL* or *PsycInfo* databases). But the bigger problem is which terminology is to be used to categorize a condition, disease, state, etc. For example, nurses often find that patient care terms are not specific enough in most medical databases while physicians don't want to use nursing terms because, well . . . they are nursing terms. Again, if a numeric or alphanumeric UCI were developed on which these other thesauri could be overlaid, it might be possible to build a metathesaurus that does not depend on the words used but strictly upon an (alpha) numeric conceptual code. This is not unlike how medical insurance claims are billed by requiring a precise code for the medical condition being treated.[1] Additionally, an (alpha) numeric system would eliminate a lot of cultural bias that exists in current thesauri where non-Western concepts don't easily fit or are untranslatable.

CONCLUSION

Both the library and scientific communities have a vested interest in bibliographic citation analysis. With the increase of non-traditional forms of electronic scholarly publishing and the recent attention brought to inaccuracies and inconsistencies in citing publications, there is further cause for these two groups to work in tandem to assure a better quality control within author reference lists and development of better electronic bibliographic linking systems so both proprietary and free access publications can be fully bibliometrically analyzed. It is interesting to speculate that if Simkin and Roychowdhury are indeed correct in their estimation that only 20% of authors read their cited articles beforehand, what does that say about the integrity of the bibliometric data that have been collected over the years? If faculty can no longer rely on the citation counts when preparing tenure and promotion packets, can institutions of higher education be convinced of focusing more on quality rather than quantity of publications and references from faculty and researchers? Is the scholarly scientific community ready to come full circle regarding citation indices so that less value is put on impact factor and citation numbers and more value is attributed to the ability of the indices to assist "the conscientious scholar to be aware of criticisms of earlier papers" (Garfield 1955).

NOTE

1. This similitude is also helpful in explaining to nurses and physicians the importance of thesauri such as the *Medical Subject Headings* (*MeSH*) in PubMed.

BIBLIOGRAPHY

Ball, P. "Paper trail reveals references go unread by citing authors." *Nature* 420 (6916): 594 (2002).

Broch, E. "Cite me, cite my references?" In *Proceedings of the 24th Annual International ACM SIGIR Conference on Research and Development in Information Retrieval*, New York: Association of Computing Machinery, 2001.

Chemistry Division. Special Libraries Association. *TriSociety–* Sunday June 9, 2002, Westin Bonaventure, Avalon Room, Program with Abstracts.* Washington, DC: Special Libraries Association, 2002 [cited 9 January 2003]. Available from World Wide Web (http://www.sla.org/division/dche/trisocrev.html).

Eichorn, P., and A. Yankauer. "Do authors check their references? A survey of accuracy of references in three public health journals." *American Journal of Public Health* 77: 1011-1012 (1987).

"Errors in citation statistics." *Nature* 415 (6868): 101 (2002).

Frazier, K. "The librarians' dilemma: Contemplating the costs of the 'Big Deal.'" *D-Lib Magazine* [online] 7 (3) (2001) [cited 14 February 2003]. Available from World Wide Web (http://www.dlib.org/dlib/march01/frazier/03frazier.html).

Garfield, E. "Citation indexes for science: A new dimension in documentation through association of ideas." *Science* 122 (3159): 108-111 (1955).

Garfield, E. *The application of citation indexing to journals management* [online]. Philadelphia: Institute for Scientific Information, 1994 [cited 8 January 2003]. Available from World Wide Web: (http://www.isinet.com/isi/hot/essays/useofcitationdatabases/9.html). First published in *Current Contents* print editions August 15, 1994.

Giles, C. L., K. D. Bollacker, and S. Lawrence. "CiteSeer: An automatic citation indexing system." In *Digital Libraries 98–Third ACM Conference on Digital Libraries*, edited by I. Witten, R. Akscyn, and F. Shipman III. New York: ACM Press, 1998 [cited 8 January 2003]. Available from World Wide Web (http://www.it-uni.sdu.dk/mmp/Library/BollackerEtAlCiteSeer99.pdf).

"Is your lab well cited?" *Nature* 227 (5255): 219 (1970).

Lascar, C., and L. D. Mendelsohn. "An analysis of journal use by structural biologists with applications for journal collection development decisions." *College & Research Libraries* 62 (5): 422-433 (2001).

Lawrence, S. "Online or Invisible?" *Nature* 411 (6837): 521 (2001).

Lawrence, S., C. L. Giles, and K. Bollacker. "Digital libraries and autonomous citation indexing." *IEEE Computer* 32 (6): 67-71 (1999).

Martino, J.P. "Research evaluation through citation indexing." In *AFOSR Research, the Current Research Program, and a Summary of Research accomplishments* (AFOSR 67-0300, AD 659-366), edited by D. Taylor. Arlington, Va.: USAF Office of Aerospace Research, 1967.

Members of the Clever Project. "Hypersearching the Web." *Scientific American.com* (June 19, 1999) [cited 9 January 2003]. Available from World Wide Web (http://www.sciam.com/print_version.cfm?articleID=000BC474-9440-1CD6-B4A8809EC588EEDF).

NEC Research Institute. *CiteSeer: Scientific Literature Digital Library*. Princeton, NJ: NEC Research Institute, Inc., 2002 [cited 9 January 2003]. Available from World Wide Web (http://citeseer.org/).

Needham, J.L. Personal correspondence. 13 February 2003.

Odlyzko, A. "The rapid evolution of scholarly communication." *Learned Publishing* 15 (1): 7-19 (2002).

Rudolph, J., and D. Brackstone. "Too many scholars ignore the basic rules of documentation." *Chronicle of Higher Education* 36 (30): A56 (1990).

Simkin, M.V., and V.P. Roychowdhury. "Read before you cite!" *ArXiv.org e-Print archive*, Cornell University: Ithaca, NY. December 2002 [cited 14 January 2003]. Available from World Wide Web (http://arxiv.org/ftp/cond-mat/papers/0212/0212043.pdf).

"Soft responses to misconduct." *Nature* 420 (6913): 253 (2002).

"SPARC/ACRL Session Gives Librarians an Economics Lesson on Serials." *Library Journal* [online]. (February 6, 2003) [cited 14 February 2003]. Available from World Wide Web (http://libraryjournal.reviewsnews.com/index.asp?layout=articlePrint&articleID=CA274516).

Stull, G. A., R.W. Christina, and S. A. Quinn. "Accuracy of references in *Research Quarterly for Exercise and Sport*." *Research Quarterly for Exercise and Sport* 62 (3): 245-248 (1991).

Sweetland, J. H. "Errors in bibliographic citations: A continuing problem." *Library Quarterly* 59: 291-304 (1989).

Sylvia, M. J. "Citation analysis as an unobtrusive method for journal collection evaluation using psychology student research bibliographies." *Collection Building* 17 (1): 20-28 (1998).

U.S. National Library of Medicine. *Unified Medical Language System (UMLS)*. Washington, DC: U.S. National Library of Medicine, 2003 [cited 13 January 2003]. Available from World Wide Web (http://www.nlm.nih.gov/research/umls/).

Walker, J. "CrossRef and SFX: complementary linking services for libraries." *New Library World* 103 (1174): 83-89 (2002).

Citation Patterns
of Advanced Undergraduate Students
in Biology, 2000-2002

Joseph R. Kraus

SUMMARY. Thirty-three undergraduate student papers in biology that were presented at an annual symposium of undergraduate research at the University of Denver from 2000 through 2002 were evaluated. There were a total of 770 citations with an average of 23.3 citations per paper. It was determined that 76.2% of the citations came from journal articles, 16.4% came from books or book chapters, 6.4% were to other miscellaneous sources, and only 1.0% were to Web sites. Other findings include the top cited journals, the oldest cited journal articles, the average age and range of books and journals, the types of miscellaneous sources cited, and the stability of the cited Web sites. *[Article copies available for a fee from The Haworth Document Delivery Service: 1-800-HAWORTH. E-mail address: <docdelivery@haworthpress.com> Website: <http://www.HaworthPress.com> © 2002 by The Haworth Press, Inc. All rights reserved.]*

KEYWORDS. Biology literature, life science literature, information-seeking patterns, bibliometric analysis, undergraduate students

Joseph R. Kraus, MLS, is Science Librarian, University of Denver, Denver, CO 80208 (E-mail: jokraus@du.edu).

The author would like to thank Jil Dawicki and the rest of the ILL staff for getting some of the obscure references in preparation for this research.

[Haworth co-indexing entry note]: "Citation Patterns of Advanced Undergraduate Students in Biology, 2000-2002." Kraus, Joseph R. Co-published simultaneously in *Science & Technology Libraries* (The Haworth Information Press, an imprint of The Haworth Press, Inc.) Vol. 22, No. 3/4, 2002, pp. 161-179; and: *Scholarly Communication in Science and Engineering Research in Higher Education* (ed: Wei Wei) The Haworth Information Press, an imprint of The Haworth Press, Inc., 2002, pp. 161-179. Single or multiple copies of this article are available for a fee from The Haworth Document Delivery Service [1-800-HAWORTH, 9:00 a.m. - 5:00 p.m. (EST). E-mail address: docdelivery@haworthpress.com].

10.1300/J122v22n03_13

INTRODUCTION

There is plenty of recent research that shows how undergraduates use the library and its print and electronic resources. However, there is little recent research that shows how undergraduate students are *citing* journals, books, and other materials in the sciences. However, there is a good amount of research that documents how graduate students and faculty in the sciences use and cite the literature. This paper will provide statistics and data to document how advanced undergraduate biology students are citing the scientific literature, and that data is compared with the citation patterns of faculty and graduate students.

Several years ago, the author of this article talked to the University of Denver Penrose Library Associate Director for Collection Development, Patricia Fisher, about his interest in citation patterns of students and faculty in the sciences. She suggested a collaborative project to evaluate some of the "honors papers" presented by University of Denver undergraduate students. Since 1996, the University of Denver has sponsored an annual Conference of Undergraduate Research. The author already knew about information usage patterns of undergraduates in the fields of physics and chemistry, but was not as familiar with the information usage patterns of life science students. After the research project was completed, the findings were presented as a poster paper at the 2000 ALA Conference in Chicago (Kraus and Fisher 2000). The following are some of the basic figures derived from that research. Over a three year span from 1997-1999, biology students wrote 51 honors papers, cited a total of 878 sources for an average of 17.2 citations per paper. Of those citations, 67% were to journals, 25% to books, 7% to miscellaneous sources, and 1% to Web sites. The average ages of book and journal citations were also found.

This paper was written to continue the data gathered from the 1997-1999 honors papers research. Because of the importance of the journal literature to biology students and faculty, it was determined that more data should be gathered concerning the top cited journal titles. In order to compare the three-year spans, the honors papers written from 2000 through 2002 were evaluated for the same basic data points in addition to determining the top cited journals.

The students who wrote honors papers received support and guidance from faculty research partners and advisors. However, the level and amount of support and guidance the students received is unknown. In order to determine that level of support, a survey of the faculty advisors may be the focus of a future research project. Based on the titles of the honors papers, and the quality of the journals that they had cited, it appears that the faculty advisors are preparing these students for graduate or medical school work. For example, here is the title to one of the student papers: "Insulin-Like Growth Factor-I (IGF-I) Offers Neuroprotection Through Suppression of Key Elements in the Intrinsic Death Signaling Pathway in Cerebellar Granule Neurons."

One may ask: "Why evaluate the citation patterns of undergraduate students? Aren't the citations going to be of low quality? Why not research the higher quality citation patterns of faculty or graduate students?" To answer, there is little research on the citation patterns of undergraduates in the sciences, and this research could fill a niche. The citation patterns of faculty can be easily gleaned by gathering the references from their published papers. One can also gather the references of Master's and Ph.D. students once the theses and dissertations are completed. Since the honors papers and references receive faculty approval before presentation at the Annual Symposium, it was deemed worthwhile to evaluate the higher quality citation patterns of those students. This paper will establish a baseline from which further research can be performed; it describes the types of material students are citing, but it does not explain how they obtain those citations.

LITERATURE REVIEW

Overview

There are many articles in the literature that demonstrate how librarians find and report citation research. Citations can be thought of as a *form* of library use, but scientists and students can get information from other sources besides libraries (Kelland and Young 1994). In other words, do not assume that a citation is a count of library use. Librarians should also be careful when using citation figures and Institute for Scientific Information (ISI) impact factors when making decisions about journals (Stankus and Rice 1982). Since undergraduate students usually do not have personal subscriptions to scholarly journals, and they have not yet formed a professional network, it is more likely that a citation is a count of library use.

Undergraduate Citation and Use Studies

In the course of gathering background literature for this paper, only a small number of prior articles that covered undergraduate citation patterns, particularly in biology, could be found (Magrill and St. Clair 1990; St. Clair and Magrill 1992). They showed that senior undergraduates in biology cited an average number of 19.2 citations per term paper. Of those citations, 75% were to journals, 14% to books, and 10% to other sources of information. They also found that biology students used older journal articles, as compared to many other disciplines. Since this research is over 10 years old, they did not evaluate how biology students cited Web resources.

More recently, researchers at Cornell University provided a thorough examination of undergraduate citation behavior in the social sciences (Davis 2002; Davis 2003; Davis and Cohen 2001). They examined the journal, book and Web citations of microeconomic term papers written by undergraduate students. In 2002, for an update article, Philip M. Davis showed that there is growth in the number of Web and newspaper citations in the students' papers. His 2003 paper documented a recent increase in the use of scholarly material because the professor provided additional guidelines on the use of scholarly and electronic resources. His research also reported a lack of persistency for cited URLs (Davis 2003). In 2001, the microeconomic students cited books 16%, scholarly journals 30%, magazines 23%, Newspapers 12%, and Web resources 13%.

Several articles document the problems of Web page change and persistence, also called "linkrot" (Koehler 1999; Koehler 2002; Lawrence, Pennock et al. 2001; Taylor and Hudson 2000). Many articles and reports also show how undergraduate students prefer to use the Internet for library resources (Friedlander 2002; Lombardo and Condic 2001; Lombardo and Miree 2003; OCLC 2002) and Internet search engines as the first place to look for research (Leibovich 2000). Research from the Pew Internet & American Life Project documents how the Internet has become central to the culture of undergraduate students (Jones 2002). Many librarians are worried that undergraduate students who first turn to the Web for research are "using unevaluated or inappropriate Web resources to support their writing assignments" (Grimes and Boening 2001).

Citation Patterns of Graduate Students in Biology

A citation analysis of biology faculty publications and graduate student papers at Temple University was performed (McCain and Bobick 1981). The researchers found that 91% of the citations were to journals. Using data from the McCain and Bobick study, Louise S. Zipp concluded that journal citations in theses and dissertations are better indicators of faculty use than had been previously assumed (Zipp 1996). She found that 70% of the top 40 faculty-cited journal titles appeared in the top 40 thesis/dissertation list of titles, and that 93% of the top 12-15 faculty-cited journal titles appeared in the top 40 thesis/dissertation list of titles.

At the University at Stony Brook, the patterns of citations in the theses and dissertations completed during the years 1989-1992 by biology graduate students were investigated (Walcott 1994). She found that graduate students studying ecology and evolution cited books three times more often than other sub-fields of biology. Journal citation rates were in the 80-95% range, depending upon the subject area. She also reported that materials published in the last ten years accounted for 80% of the citations.

Citation Patterns of Faculty/Professionals

Recently, it was found that 91.3% of the total citations from molecular biology faculty at the University of Illinois Chicago were to journal articles (Hurd, Blecic et al. 1999). They also noted that "molecular biology is a field with a high degree of immediacy where the most current journals are likely to be the most heavily used." They surmised that electronic sources of information will be important in the future, but they did not see any current citations to specific Web sites. "None of the forty-four articles in the sample of faculty publications cited any electronic resource . . . The authors expect that references to gene databanks will be found increasingly in their publications."

Faculty at the School of Life Sciences at the University of Illinois at Urbana-Champaign cited an average of eighteen citations per article (Schmidt, Davis et al. 1994). A total of 83% of all citations were to journals. They also found that 29% of the journal titles accounted for 80% of the journal citations. A citation analysis of publications from the faculty of the Department of Environmental and Human Health was performed at Texas Tech University (Johnson 2002). He found that "the average age of citations was 10.5 years for journals and 9.4 years for books. On average, journals were cited 67% of the time, while books were cited 17% of the time." Citations of biology faculty were analyzed at Washington State University (Crotteau 1997). He listed the top 100 cited journals. Faculty citations were compared to journal holdings, and he surveyed the faculty to learn how they obtained cited items. On the lower end of the scale, it was found that water resources researchers cited journals only 54% of the time (Walker and Ahn 1995).

Librarians also use citation information as supporting data during a cancellation project or to justify subscriptions. A serials review was undertaken in 1993 at Pennsylvania State University (PSU) Pattee (Life

Sciences) Library to ascertain a core list of journals in molecular and cellular biology (Hughes 1995). She created an interesting point system for determining the importance of a journal to the local collection. A journal received 5 points if a PSU author published an article in the journal, and 1 point if a PSU author cited the journal. As part of a cancellation project at the Falconer Biology Library at Stanford University, a librarian performed a comparative analysis of citation studies, in-house use, and ISI Impact Factor for determining which journals had the highest local use (Wible 1989). A researcher at the Marine Science Research Center at SUNY-Stony Brook created a list of core journals compiled from journal titles in which 60 local scientists published articles (Williams 1989). In addition, a list of cited journal titles was collected. These two lists were used to justify expenditures, and a comparison was made with the ISI core journals in oceanography and marine biology.

Citation Analysis for Specific Journals

In some cases, citation statistics for specific journals in the biological sciences were evaluated. The bibliometric properties of two journals in plant biology–the *Canadian Journal of Plant Science*, and the *Canadian Journal of Botany*–were analyzed (Nordstrom 1987). The author found that articles in the two journals cited journal articles at different rates, 68% and 77% respectively. A citation analysis of three systematic botany journals, *Brittonia*, *Systematic Botany*, and *Taxon* was performed (Delendick 1990). He found that 66% of the citations from those three journals were to journal articles. He also provided an analysis of the age of the cited material and a ranking of the most highly cited journals.

METHODOLOGY

The research presented here was undertaken in an effort to better understand the citation patterns of advanced undergraduate students in biology. After reading some of the prior literature listed above, it was possible to see the types of data other researchers collected. It was desired to collect the same types of citation data for easier comparison. The following questions were developed:

1. What was the average number and range of citations per student paper?
2. What percentage of students' citations were to books or book chapters?

3. What percentage of students' citations were to journals or magazines?
4. What percentage of students' citations were to other miscellaneous resources? What kinds of other resources were those?
5. What percentage of students' citations were to Internet resources? Are those citations still valid?
6. What were the average ages of cited journals and books? What was the range?
7. What were the top cited journals?
8. What older journals were cited?

Once the questions were finalized, then it was possible to know what data points to collect for the research.

The references listed at the end of 33 honors papers presented in 2000 through 2002 were photocopied. There were 10 papers from 2000, 10 from 2001, and 13 from 2002. From those 33 papers, a total of 770 references were counted. Each citation was marked as either a journal article, a book or a book chapter, an "other" miscellaneous source, or a Web citation. Using the method Chandra G. Prabha developed for determining journal article characteristics (Prabha 1996), citations from material such as annuals, monographic series, newspapers and other irregular serials were not counted as journal articles. If a citation to a periodical (such as an annual review) was located in our book stacks, then the citation was counted as a book cite. A small number of government documents were counted as "other" miscellaneous and Web sources. In a small number of cases, it was difficult to determine if a citation was to a book or to a journal, and in those cases, the Colorado Unified Catalog-Prospector (http://prospector.coalliance.org), was used to help determine the nature of the material. For example, if the University of Colorado at Boulder had an item in their book stacks, then the citation was counted as a book cite.

To calculate the age of citations, the difference between the year of the student paper, and the year of the cited publication was found. For example, when a student wrote a paper in 2002, and cited an article from the *Journal of Whatever* from 1989, the calculated the age of the citation was marked as thirteen years. To calculate the average age of journal citations for each student paper, the ages of citations for the journals were added up and divided by the number of cited journals. The same methodology was used for calculating the average age of cited books.

RESULTS AND DISCUSSION

General Citation Statistics

From the 33 student papers, there were a total of 770 citations for an average of 23.3 citations per paper. One student paper only had three citations, while another student paper had eighty-four citations. As expected, most of the citations, 587 out of 770, were to journal articles. There were 126 book citations, 49 other miscellaneous citations, and only eight Web citations. Six of the eight Web citations came from a single student paper on stem cell research. See Figure 1 and Figure 2 for the general statistics. Figure 3 shows the average number of citation distribution for each publication format.

These general figures are not much different from the numbers gathered in 2000 (Kraus and Fisher 2000). However, citations to Web sites did not go up as expected. Web citations were only 1% for the study conducted on papers done in 1997-1999, and for papers done in 2000-2002. Had it not been for the one paper on stem cell research, the percentage of Web site citations would have gone down.

Book Citations

Twenty-eight of the 33 papers contained citations to books or book chapters. As shown in Figure 4, most of the citations to book material were 25 years or less, and the average age of a book citation was 19.13 years (see Figure 5). However, there were two papers that cited many older books; several of their citations were to books that were well over 100 years old. These two papers contained book citations that averaged over 47 years. If those two papers are ignored as statistical outliers, the aver-

FIGURE 1. Citation Statistics

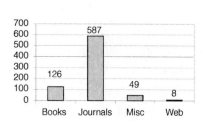

FIGURE 2. Basic Citation Data, Pie Chart

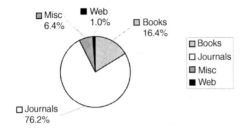

FIGURE 3. Average Number of Citations per Paper by Format

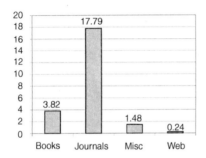

age age of all book citations drops down to 11.36 years. This result is a closer match to the data found in the 2000 study (Kraus and Fisher 2000).

Journal Citations

All 33 of the student papers contained citations to journal articles. As shown in Figure 6, most of the student papers cited journal articles that averaged under 18 years. The average age of a journal citation was 10.65 years old (see Figure 5). Table 1 provides a list of the top cited journals; 46 journals were cited three or more times. The University of Denver has a subscription to most of those top cited journals. Of those, faculty and students have electronic access to most. Some of those publications were cancelled some time ago, such as *Brain Research*. Some journal titles have older articles freely available on the HighWire Press

FIGURE 4. Sorted Average Age of Book Citations

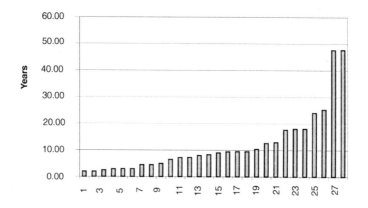

FIGURE 5. Average Ages of Citations by Format

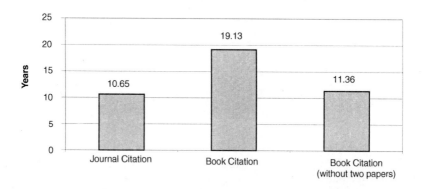

Web site (http://highwire.stanford.edu), such as the *EMBO Journal*. Even though the University of Denver does not have a current subscription to the *EMBO Journal*, students and faculty can easily retrieve older scientific articles.

For Table 2, journal titles are listed by the age of the oldest journal article citation. A natural break-off point appeared at 40 years. Another 17 journal articles were cited between 30 and 40 years old; those were from journals such as: *Journal of Cell Biology, Journal of Biological Chemistry, Science*, and *Endocrinology*.

FIGURE 6. Sorted Average Age of Journal Citations

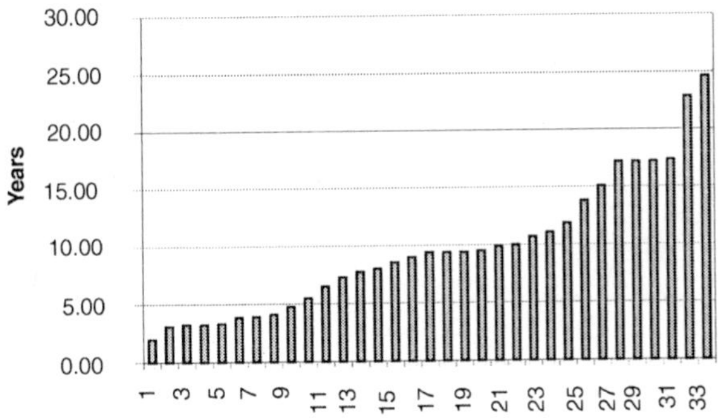

Other Miscellaneous Citations

Table 3 lists some of the other miscellaneous citations students used. Ten of the thirteen "unpublished manuscripts" were cited by one student paper. All seven of the newspaper article citations came from a single paper on stem cell research.

Web Citations

Since only 1% of the citations were to Web sites, there were not that many to evaluate. From the data presented in Table 4, one can see that five of the eight were perfectly cited. Two of the eight citations were indirect. The student provided the base URL for the organization publishing the article, but the rest of the URL was not provided for a direct link. One had to search the Web site to find the cited article title. Only one of the eight Web citations was a dead link. After performing a Web search, it was possible to find the same directory structure (and probably the same content) for the publication at another Web site.

Other Discussion

There were 13 faculty advisors over the three-year span, and the papers tended to focus on topics of interest to the faculty members. The topics of the papers appeared to be closely related from one year to the next. From looking at the top cited journals, it is easy to see that many of the students did work in endocrinology.

TABLE 1. Top Cited Journals, Including Ties

Rank	Journal Title	Number of Citations
1	Journal of Cell Biology	35
2	Nature	31
3	Proceedings of the National Academy of Sciences of the United States of America (PNAS)	30
4	Journal of Biological Chemistry	22
5	General and Comparative Endocrinology	19
6	Science	16
7	Cell	13
8	Cell Motility and the Cytoskeleton	12
8	Endocrinology	12
10	Chromosoma	11
11	Journal of Neuroscience	10
12	Development	9
13	Developmental Biology	8
13	Journal of Comparative Neurology	8
13	Nucleic Acids Research	8
16	Peptides	7
17	Behavioral Ecology and Sociobiology	6
17	Biochemical and Biophysical Research Communications	6
17	Neuron	6
20	Archives of Histology and Cytology	5
20	Current Opinion in Cell Biology	5
20	New England Journal of Medicine	5
23	Brain Research	4
23	Endocrine Reviews	4
23	Environmental Science and Technology	4
23	Experimental Neurology	4
23	FEBS Letters	4
23	Journal of Cell Science	4
23	Journal of Molecular Biology	4
23	Molecular Endocrinology	4
23	Molecular and Cellular Biology	4
23	Molecular Biology of the Cell	4
23	Neuroendocrinology	4
23	Physiological Reviews	4
23	Trends in Cell Biology	4
36	Biochimica et Biophysica Acta	3
36	Bioscience	3
36	Cancer Research	3

Rank	Journal Title	Number of Citations
36	Diabetes	3
36	EMBO Journal	3
36	Federation Proceedings (FASEB)	3
36	Genes and Development	3
36	Journal of Experimental Zoology	3
36	Journal of Molecular Evolution	3
36	Proceedings of the Royal Society: Biological Sciences	3
36	Radiocarbon	3

TABLE 2. Journal Titles Sorted by the Age of the Oldest Cited Article Within the Journal

Journal Title	Years Old
Journal of Comparative Neurology and Psychology	94
American Journal of Psychology	89
Paeobiologica	72
Journal of Comparative Neurology and Psychology	70
Nature	64
American Journal of Anatomy	59
American Journal of Anatomy	59
Psychological Review	54
Physiological Review	52
Journal of Biological Chemistry	51
Bulletin of the Los Angeles Neurological Society	50
Ecology	46
Psychological Review	46
Canadian Journal of Biochemistry and Physiology	45
Endocrinology	45
Experimental Cell Research	41
Pacific Insects	41

If an article was cited as "in press" or submitted to the *Journal of Whatever*, Biological Abstracts or Medline was used to determine where and when the paper was finally published. If the citation could be verified as a published journal article, that citation was counted as a journal article and the year of publication was noted. No student cited an article that was published in a year after the student paper was completed.

Typos and citation mistakes were noticed, such as the wrong article title, wrong journal name, wrong volume number, no year listed, and

TABLE 3. Listing of Miscellaneous Citations

Types of Miscellaneous Citations	Amount
Unpublished manuscript/data	13
Prior honors paper	9
Newspaper article	7
Technical report	4
Lecture notes/discussion	3
Personal communication	3
Software and software documentation	3
Government document	2
Conference report/abstract	2
Bad citation (impossible to get more info.)	1
Ph.D. thesis	1
Laboratory manual	1
TOTAL	49

TABLE 4. Listing of Cited Web Sites

URL	Notes
www.jlaw.com (Article cited - "Stem Cell Research in Jewish Law")	Indirect citation. A direct citation to http://www.jlaw.com/Articles/stemcellres.html would have been better.
www.americancatholic.org (Article cited -"Is stem cell research moral?")	Indirect citation. A direct citation to http://www.americancatholic.org/News/StemCell/ask_stemcell.asp would have been better.
employeeweb.myxa.com/rrb/Audobon/VolIV/00457.html	Does not work. It appears as if this Web site had "migrated" over to: http://www.abirdshome.com/Audubon/VolIV/00457.html, digitized by Richard R. Buonanno, probably the "rrb" formerly with myxa.com.
www.unesco.org/ibc/en/reports/embryonic_ibc_report.pdf	Good.
www.nih.gov/news/stemcell/primer.htm	Good.
www.nih.gov/news/stemcell/stemcellguidelines.htm	Good.
escr.nih.gov	Good.
Cellbio.utmb.edu/cellbio/microtubule_structure.htm #hydrolysis	Good.

others. For the most part, the data as presented in the references section of the paper were used. If a student accidentally indicated the wrong year for a publication, that wrong year was used for the age calculation. If the student forgot to mark down the year of the article, then an indexing service was used to verify the year. Judith A. Harper found a great deal of citation error in two volumes of the journal, *Environmental and Experimental Botany* (Harper 2001). While the data for this project was not checked for citation errors, it is likely that some of the students had cited the wrong year for journal article citations, and those errors may have introduced errors into this research.

Looking at some of the citations, it appears some of the honors students had used the personal library, journal login and passwords of their faculty advisor. Some of the citations noted Web access to the journal *Nature*. Since the University Library did not have an electronic subscription to the journal at the time, the student may have used a personal Web subscription from the faculty advisor, or the student may have gone to another institution for electronic access. In such cases, the student cited an electronic document that was not available through a campus license and was not retrieved through interlibrary loan. Many of the students cited publications that were authored by their faculty advisor. It is possible the faculty members handed them articles from their personal collection as background reading material. Thus, the students may have been using the personal library of their advisor instead of the University Library to find some of their research articles.

Some of this data could be used for collection development purposes. The library does have paper subscriptions to the *Journal of Cell Biology* and the *Journal of Biological Chemistry*, but we have been hesitant to pay extra for electronic access to these two journals. Since the data indicates high local use, electronic subscriptions could be explored to supplement or replace the paper subscriptions.

CONCLUSION

Many of the citation figures agree pretty well with prior research. The students cited journal literature at a much higher rate than books or other sources. Faculty and students may be using electronic databases and electronic journals for their research, but they do not cite very many "free" Web sites in their papers. This is consistent with the findings of Hurd, Blecic and Vishwanatham (1999), but it does not agree very well with the findings of Davis (2003) at Cornell University.

A reason could be the faculty advisors at the University of Denver had greater influence on the students than the microeconomics faculty at Cornell on their students. Without the close advisement of the teaching faculty, the biology students may have cited more Web resources. It is well documented that science faculty and students prefer electronic access to academic sources of information (Brown 1999; Brown 2001; Curtis, Weller et al. 1997; Hurd 2000; Hurd 2001; Morse and Clintworth 2000; Quigley, Peck et al. 2002; Rogers 2001). The University of Denver students are certainly using electronic resources at a higher rate than print resources to get abstracts and full-text articles, but they are not citing the use of the Web as the delivery mechanism, or Web sites for very many sources of scholarly information. Similar to the findings of Lisa M. Covi, it is likely that the undergraduate students used electronic resources to "mimic" the electronic research patterns of their advisor (Covi 2000). While many science faculty use electronic resources to access scholarly materials, many do not trust the accuracy and reliability of "free" Web content (Herring 2001). Further research may be undertaken to determine how the students obtained their research materials, and the nature and extent of the faculty advisement.

REFERENCES

Brown, Cecelia M. 1999. Information seeking behavior of scientists in the electronic information age: astronomers, chemists, mathematicians, and physicists. *Journal of the American Society for Information Science* 50(10): 929-43.

Brown, Cecelia M. 2001. The E-volution of preprints in the scholarly communication of physicists and astronomers. *Journal of the American Society for Information Science and Technology* 52(3): 187-200.

Covi, L. M. (2000). "Debunking the myth of the Nintendo generation: How doctoral students introduce new electronic communication practices into university research." *Journal of the American Society for Information Science* 51(14): 1284-1294.

Crotteau, Mark. 1997. Support for biological research by an academic library: A journal citation study. *Science & Technology Libraries* 17(1): 67-86.

Curtis, Karen L., Ann C. Weller and Julie M. Hurd. 1997. Information-seeking behavior of health sciences faculty: the impact of new information technologies. *Bulletin of the Medical Library Association* 85(4): 402-10.

Davis, Philip M. 2002. The Effect of the Web on undergraduate citation behavior: A 2000 update. *College & Research Libraries* 63(1): 53-60.

Davis, Philip M. 2003. The Effect of the Web on undergraduate citation behavior: Guiding student scholarship in a networked age. *Portal: Libraries and the Academy* 3(1): 41-51.

Davis, Philip M. and Suzanne A. Cohen. 2001. The effect of the Web on undergraduate citation behavior 1996-99. *Journal of the American Society for Information Science and Technology* 52(4): 309-314.

Delendick, Thomas J. 1990. Citation analysis of the literature of systematic botany: a preliminary survey. *Journal of the American Society for Information Science* 41(7): 535-543.

Friedlander, Amy. 2002. *Dimensions and Use of the Scholarly Information Environment: Introduction to a Data Set Assembled by the Digital Library Federation and Outsell, Inc.* Washington, D.C., Digital Library Federation and Council on Library and Information Resources. Available at http://www.clir.org/pubs/reports/pub110/contents.html.

Grimes, Deborah J. and Carl H. Boening. 2001. Worries with the Web: A look at student use of Web resources. *College & Research Libraries* 62(1): 11-23.

Harper, Judith A. 2001. Citation inaccuracy in a scientific journal: A continuing issue. *Science & Technology Libraries* 20(4): 39-44.

Herring, Susan D. 2001. Faculty acceptance of the World Wide Web for student research. *College & Research Libraries* 62(3): 251-258.

Hughes, Janet A. 1995. Use of faculty publication lists and ISI citation data to identify a core list of journals with local importance. *Library Acquisitions* 19: 403-13.

Hurd, Julie M. 2000. The transformation of scientific communication: A model for 2020. *Journal of the American Society for Information Science* 51(14): 1279-1283.

Hurd, Julie M. 2001. Digital collections: Acceptance and use in a research community. *Crossing the Divide: Proceedings of the Tenth National Conference of the Association of College and Research Libraries, March 15-18, 2001, Denver, Colorado.* H. A. Thompson. Denver, CO, ACRL: 312-319. Available at http://www.ala.org/acrl/papers01/hurd.pdf.

Hurd, Julie M., Deborah D. Blecic, and Rama Vishwanatham I. 1999. Information use by molecular biologists: implications for library collections and services. *College & Research Libraries* 60(1): 31-43.

Johnson, William T. 2002. Environmental impact: A preliminary citation analysis of local faculty in a new academic program in environmental and human health applied to collection development at Texas Tech University Library. *LIBRES: Library and Information Science Research Electronic Journal* 9(1). Available at http://libres.curtin.edu.au/libre9n1/toxcite.htm.

Jones, Steve. 2002. *The Internet Goes to College: How Students are Living in the Future with Today's Technology.* Washington, DC, Pew Research Center, Pew Internet and American Life Project: 1-22. Available at http://www.pewinternet.org/reports/toc.asp?Report=71.

Kelland, John Laurence and Arthur P. Young. 1994. Citation as a form of library use. *Collection Management* 19(1/2): 81-100.

Koehler, Wallace. 1999. An analysis of Web page and Web site constancy and permanence. *Journal of the American Society for Information Science* 50(2): 162-80.

Koehler, Wallace. 2002. Web page change and persistence–a four-year longitudinal study. *Journal of the American Society for Information Science and Technology* 53(2): 162-71.

Kraus, Joseph R. and Patricia Fisher. 2000. *Citation Analysis of Undergraduate Biology Department Honors Papers at the University of Denver.* ALA Conference, Chicago, ALA. Available at http://www.du.edu/~jokraus/ALA2000/biocites.html.

Lawrence, Steve, David M. Pennock et al. 2001. Persistence of Web references in scientific research. *Computer* 34(2): 26-31.

Leibovich, Lori. 2000. Choosing quick hits over the card catalog: Many students prefer the chaos of the Web to the drudgery of the library. *New York Times.* New York: August 10, 2000. G1.

Lombardo, Shawn V. and Kristine S. Condic. 2001. Convenience or content: a study of undergraduate periodical use. *Reference Services Review* 29(4): 327-337.

Lombardo, Shawn V. and Cynthia E. Miree. 2003. Caught in the Web: The impact of library instruction on business students' perceptions and use of print and online resources. *College & Research Libraries* 64(1): 6-22.

Magrill, Rose Mary and Gloriana St. Clair. 1990. Undergraduate term paper citation patterns by disciplines and level of course. *Collection Management* 12(3-4): 25-56.

McCain, Katherine W. and James E. Bobick. 1981. Patterns of journal use in a departmental library: A citation analysis. *Journal of the American Society for Information Science* 32(4): 257-67.

Morse, David H. and William A. Clintworth. 2000. Comparing patterns of print and electronic journal use in an academic health science library. *Issues in Science and Technology Librarianship*, Issue 28 (Fall). Available at http://www.library.ucsb.edu/istl/00-fall/refereed.html.

Nordstrom, L. O. 1987. Applied versus basic science in the literature of plant biology: A bibliometric perspective. *Scientometrics* 12(5-6): 381-393.

OCLC White Paper on the Information Habits of College Students. 2002. *How Academic Librarians Can Influence Students' Web-Based Information Choices*. Dublin, Ohio, OCLC. Available at http://www2.oclc.org/oclc/pdf/printondemand/informationhabits.pdf.

Quigley, J., D. R. Peck, et al. 2002. Making choices: factors in the selection of information resources among science faculty at the University of Michigan results of a survey conducted July-September, 2000. *Issues in Science & Technology Librarianship* Issue 34. Available at http://www.istl.org/02-spring/refereed.html.

Prabha, Chandra G. 1996. *Characteristics of Articles Requested through OCLC Interlibrary Loan*. Dublin, OH, OCLC. Available at http://www.oclc.org/research/publications/arr/1996/artill.htm.

Rogers, Sally A. 2001. Electronic journal usage at Ohio State University. *College & Research Libraries* 62(1): 25-34.

Schmidt, Diane, Elisabeth B. Davis et al. 1994. Biology journal use at an academic library: a comparison of use studies. *Serials Review* 20(2): 45-64.

St. Clair, Gloriana and Rose Mary Magrill. 1992. Undergraduate use of four library collections: format and age of materials. *Collection Building* 11(4): 2-15.

Stankus, Tony and Barbara Rice. 1982. Handle with care: Use and citation data for science journal management. *Collection Management* 4(1/2): 95-110.

Taylor, Mary K. and Diane Hudson. 2000. "Linkrot" and the usefulness of Web site bibliographies. *Reference & User Services Quarterly* 39(3): 273-277.

Walcott, Rosalind. 1994. Local citation studies–A shortcut to local knowledge. *Science & Technology Libraries* 14(3): 1-14.

Walker, Richard D. and Myeonghee Lee Ahn. 1995. The literature cited by water resources researchers. *Changing Gateways: The Impact of Technology on Geoscience Information Exchange–Geoscience Information Society. Meeting (29th:1994: Seattle, Wash.)*. B. E. Haner and J. O'Donnell, Geoscience Information Society: 67-77.

Wible, Joseph G. 1989. Comparative analysis of citation studies, swept use, and ISI's impact factors as tools for journal deselection. *IAMSLIC at a crossroads–Interna-*

tional Association of Marine Science Libraries and Information Centers. Conference (15th: 1989: St George's, Bermuda). R. W. Burkhart, International Association of Marine Science Libraries and Information Centers: 109-116.

Williams, D. C. (1989). Using core journals to justify subscriptions and services. *IAMSLIC at a crossroads–International Association of Marine Science Libraries and Information Centers. Conference (15th: 1989: St George's, Bermuda).* R. W. Burkhart, International Association of Marine Science Libraries and Information Centers: 123-134.

Zipp, Louise S. 1996. Thesis and dissertation citations as indicators of faculty research use of university library journal collections. *Library Resources & Technical Services* 40(October): 335-342.

Index

Page numbers followed by t indicate tables; those followed by f indicate figures.

Abel, R., 89
Abridged Index Medicus, 94
Access Impact, 108-109
ACLS History Project, 32
ACM. *See* Association for Computing
    Machinery (ACM)
ACRL. *See* Association of College and
    Research Libraries (ACRL)
ACS. *See* American Chemical Society
    (ACS)
ACS Journal Archives, 129
ACS Web Editions, 129,130
Agre, P.E., 78
ALA. *See* American Library Association
    (ALA)
Allen, R.S., 44
Amazon.com's book recommender, 94
American Chemical Society (ACS),
    31,60,122
American Institute of Physics, 128,130
American Library Association (ALA),
    102,155
American Mathematical Society, 16,130
American Physical Society (APS), 60,122
Annual Association of American
        Publishers Professional and
        Scholarly Publishing
        Conference (February 2003),
        152
APS. *See* American Physical Society
    (APS)
Archive(s), pre-print, in scientific and
    technical journals, 22
Archiving

electronic
    defined, 118-123
    pricing models for, 129-130
    of electronic content, 126-128
    of electronic journals, issues related
        to, 113-136. *See also*
        Electronic journals, archiving
        of, issues related to
    who should do, 123-126
Archiving Policies Survey, 135-136
ARLs. *See* Association of Research
    Libraries (ARLs)
Armed Services Technical Information
    Agency (ASTIA), 83
Arms, 92
arXiv, role of, 10-13
Aslib Cranfield Project, 83
Association for Computing Machinery
    (ACM), 127
Association of College and Research
    Libraries (ACRLs), 155
Association of Research Libraries
    (ARLs), 24,41,102
ASTIA. *See* Armed Services Technical
    Information Agency (ASTIA)
Attitude(s), of online researchers,
    factors in changing, 140-143

Baily, C.W., 23
Barschall, H.H., 40,41-42,60
Behavior(s), of online researchers,
    factors in changing, 140-143

181